SDGs and Textiles

Series Editor
Hafeezullah Memon, International Institute of Silk, Zhejiang Sci-Tech University, Hangzhou, China

The book series "SDGs and Textiles" addresses the strategies to achieve sustainable development goals (SDGs) in the present, past, and future. It presents books about the present and future policies of textile ministries of different countries, and books related to sustainability education around different parts of the world in the textile sector. Moreover, it would welcome the conference proceeding related to SDGs and Textiles. The series would cover books comparing the sustainability and SDGs of different institutions and countries. The individual book volumes in the series are thematic. The goal of each book is to give readers a comprehensive overview of a different area of sustainability in the textile sector. As a collection, the series provides valuable resources to a broad audience in academia, the research community, industry, and anyone looking to expand their knowledge of SDGs and Textiles.

Textiles and life are together – life cannot be separated from textiles as it is the most important need for human beings after food. In 2015, the United Nations General Assembly proposed 17 interlinked global goals to be achieved by 2030. Since then, academia and industry have paid much attention to achieving these goals. Textile found its close relation with almost all of these 17 goals.

SDG 1 - No Poverty: Poverty would never be overcome by a charity only; it is essential to develop people's skills to have a better and wealthy life. Thus, the textile can be considered an excellent discipline to achieve this goal by creating jobs and small and medium businesses.

SDG 2 - Zero Hunger: Through the effective utilization of advanced application of Agrotech Textiles, it is possible to have higher crop yields and save crops from rough weather, unexpected rains, floods, insects, etc.; thus, geotextiles play an essential in achieving this goal of sustainable development.

SDG 3 - Good Health & well-being: There has been much health consciousness after Covid19, and medical textiles assist in getting good health and well-being.

SDG 4 - Learning & Education: Textile or fashion has remained a significant discipline for societies for ages, and there has always remained much to explore in this field. Textile-related universities may play a vital role by offering free access to their education resources, training and spreading information among the locals.

SDG 5 - Gender Equality: The textile sector is one of the industrial sectors that accepted gender equality long ago; in particular, the garment sector has more females than males. Thus, the textile sector has been doing gender equality. Moreover, there has been a recent trend for Gender Neutral Clothing, which need worth studying and may further assist gender equality.

SDG 6 - Clean Water & Sanitation: Textiles could be achieved through filtration, and of course, textile is one of the critical materials for filtration.

SDG 7 - Affordable & Clean Energy: With the recent advancement in material science and engineering, the textile sector has come on the front for, not only by using this clean energy during textile production but also by assisting the production of this clean energy, either in the form of wind turbines blades made of textile composites or by energy harvesting from T-Shirts, etc.

SDG 8 - Decent Work: Recently, there has been much attention that the textile workers are not paid well, labor rights are not cared about, etc.

SDG 9 - Industry and innovation: Textile Industry always follows innovation; the textile companies that do not chase innovation cannot survive in the market.

SDG 10 - Reduced Inequalities: Getting better life and well-being would help reduce inequalities in the textile industry.

SDG 11 - Sustainable Cities: Sustainable Textile Cities through Buildtech and transport textiles.

SDG 12 - Consumption and Production: Textile and garment consumption and production all come under.

SDG 13 - Climate Action: Oekotech or Ecotech Textile, waste management of textiles are upfront to achieve this goal of sustainable development.

SDG 14 - Life Below Water: Mitigating microfiber waste in rivers and oceans may come under the context of it. There has been much attention on this subject after passing the bill at the parliament level of the UK.

SDG 15 - Life on Land: Geotech or Geotextiles studies life on land.

SDG 16 - Peace, Justice, and Strong Institutions: Protective textiles are doing their best to achieve peace, justice, and strong institutions.

SDG 17 - Partnerships for the Goals: The application of textiles to achieve sustainable development goals is only an example. In all textiles sectors, combined efforts of all the goals are essential to achieve true sustainability.

Hafsa Jamshaid · Usman Zubair
Editors

Sustainable Recycling of Fiber Reinforced Polymer Composites

Springer

Editors
Hafsa Jamshaid
School of Engineering & Technology
National Textile University
Faisalabad, Pakistan

Usman Zubair
School of Engineering & Technology
National Textile University
Faisalabad, Pakistan

ISSN 2948-1236 ISSN 2948-1244 (electronic)
SDGs and Textiles
ISBN 978-981-96-9535-5 ISBN 978-981-96-9536-2 (eBook)
https://doi.org/10.1007/978-981-96-9536-2

This Springer imprint is published by the registered company Springer Nature Singapore Pte Ltd.
The registered company address is: 152 Beach Road, #21-01/04 Gateway East, Singapore 189721, Singapore

Contents

Overview of the Recycled Fiber Reinforced Polymer Composites

Hafsa Jamshaid

Abstract Recycled Fiber Reinforced Polymer Composites (FRPC) offer **an exciting** opportunity for sustainable innovation, indicating a greener future for industries ranging from construction to automotive manufacturing. Recycling of FRPC helps to develop the relationship between sustainability and innovation, offering a tangible solution to the pressing environmental challenges of the present time. With their wide-ranging applications and eco-conscious manufacturing processes, they stand as a beacon of hope for industries striving to embrace greener practices and mitigate their ecological impact. This chapter explains composites and their importance in various sectors. Discussion about composite materials' environmental consequences, the necessity for sustainable management, and the importance of sustainable materials is discussed. The chapter explores the overview of recycled fiber sources, their properties, need of recycling composites, their manufacturing techniques, and performance evaluation.

Keywords Fiber reinforced composites · Recycling · Reuse · Applications · Circular economy

1 Introduction

The art of textiles has been an integral part of our lives since ages. The selection of farbric materials in specified applications from wearables to industrial usage, is linked with raw materials for specific properties. There are two main classifications of fibers. They can be extracted from natural or manmade origin. Natural fibers are sourced from plants, animals and minerals, while man-made fibers are synthetically produced either from petrochemical or from natural source. Both types of fibers have many uses in our everyday life and their demand is continuously increasing due to increase in population, fast fashion and change of lifestyles. But the environmental implications

H. Jamshaid (✉)
School of Engineering and Technology, National Textile University Faisalabad, Faisalabad, Pakistan
e-mail: hafsa@ntu.edu.pk

H. Jamshaid and U. Zubair (eds.), *Sustainable Recycling of Fiber Reinforced Polymer Composites*, SDGs and Textiles, https://doi.org/10.1007/978-981-96-9536-2_1

of their production and disposal after usage have raised significant concerns. Textile waste management is increasingly important due to the environmental footprint of textiles. There are two broad classifications of waste depending on the source, i. e post-consumer and Pre-consumer.

Post-consumer waste is the one which is discarded by consumers like old clothes while Pre -consumer waste are generated during manufacturing those can be either defected or production waste. As technology is changing at a fast pace the requirement of materials for different applications is also changing rapidly. Modern era products are presented with multiple specifications and properties that are not possible to achieve with common materials i.e metal materials. Materials for construction and transportation industries are specially demanding in terms of properties. Metal alloys are replaced by composites in multiple applications such as aerospace applications, construction automobiles, sports goods, marine etc. Composites can be defined as a combination of dissimilar materials to perform tasks that neither of the constituent materials can perform alone, since hardly any material is used in its pure form today, have all-rounder properties. Composite materials' history is as old as mankind itself. The beginning of composites is not defined but every data present in records suggest some forms of materials were used as composites; the most probable first ever composite was adobe brick where mud was used as the matrix and grass or straw was used as the preform. Composite is simply a material made up of two or more than two materials of dissimilar physical or chemical properties. Different types of composites exist in this world; those can be classified based on their application as mentioned in Fig. 1. In modern life, the wide range of applications of these materials is evident. The reinforcement is impregnated in a matrix that can be different depending on the application area. In the present era most of the research & development are on textile reinforced composites. They are used in various applications of technical textiles from sports equipment's to aerospace [1]. The scope of this book is Textile reinforced composites (TRC), so further discussion will be related to TRC. The textile composites are composed of two materials, i.e., a textile skeleton for reinforcement (called preform) and a binding adhesive (called matrix) material to keep the skeleton integrated into a specific shape. In the modern era, composites formulated by textile reinforcement are majorly in use. The scope of their usage is ever increasing across the globe based on performance and application being on offer. While making a comparison with conventional materials, composites with fibers exhibit multiple advantages such as small specific weight, low factor of heat conductivity, low cost of design installation, corrosion resistance, high operating temperature, chemical inertness, long wearing life, high specific mechanical properties. For improvement in mechanical properties of the fiber reinforced composites, the most common fiber for reinforcement is glass fiber. Other fibers which are used for reinforcement are plastic and carbon fibers.

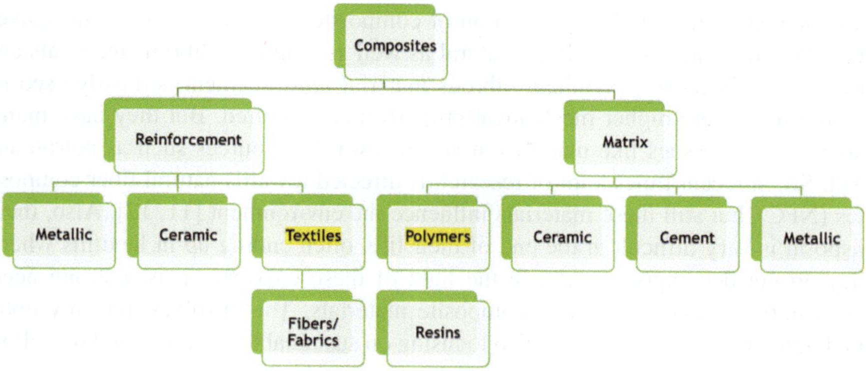

Fig. 1 Classification of composites

2 Market Share of Composites and Growing Opportunities

The demand of composites and composite-based products has increased at a very high pace. FRPCs are developing quickly due to aerospace, construction, and transportation demand [3] There is an expected growth in global composite markets with an annual CAGR of 10.8% and is projected to reach USD 181.7 billion by 2028 from USD 108.8 billion in 2023 [2]. The Textile Composites market is witnessing significant growth. It is expected to register a CAGR of greater than 6% from 2025 to 2030 [3]. As global composite end product business has grown dramatically, with aviation rising to \$8.9 billion in six years, building to \$14.7 billion, and transportation to \$22.4 billion [4–6]. The commonly used composites are fiber glass and polyester composites. Until 2013 USA was the biggest consumer. Now a days China has become the largest consumer of composites. Transportation and automobile are the main industries where contribution of composites has gained very high usage. Composites have made a significant contribution in the aerospace industry. There is considerable increase of composites in the aerospace industry and an overview explains that it has been increased up to 10.3% even higher than automotive industry. For example, the new generation aircraft Boeing 787 and Air Bus 350 developed with up to 50% of carbon fiber composites. The aerospace composites market size was valued at USD 25.91 billion in 2023. The market is anticipated to grow from USD 29.06 billion in 2024 to USD 73.36 billion by 2032, exhibiting a CAGR of 12.3% during 2024–2032 [5–9]. The construction industry accounts for 42% of all composite material usage, making it the largest market for composites among all other industries. In this industry the share of composite materials has risen. The composites share has increased significantly up to 4.9%. Pipes and tanks are another major industry where the utilization of composites has increased up to 5.2%. the major consumption of pipes and tanks are in the transportation of oil and gas, fuel, waste water and sewerage etc. [9]. Further utilization and adaptation of composites is prone to limitation due to waste managament issue. Their widespread use has led to concerns regarding their environmental impact. The production of composite materials is energy intensive, required raw materials which are natural as well as manmade fibrous materials are used as reinforcement. Synthetic fibrous material reinforcement is mostly used in composites where higher mechanical properties are required. But they have many environmental issues like made from non-renewable resources such as petroleum [10]. So, in recent times a lot of research is directed towards natural fiber composites (NFC), but still these materials influence the environment [11, 12]. Also, their disposal is very difficult at the end of their life, often ending up in landfills where they do not decompose easily. In the light of these concerns, it is a strong need for sustainable management of composite materials. This involves not only optimizing their design and use but also focusing on sustainable materials and recycling technologies.

3 Significance of Recycled Fiber-Reinforced Polymer Composites

Recycling fiber and composite materials are gaining significant attention due to their environmental benefits and the growing need to reduce waste. Composites are the products which are specialty oriented in terms of end use and addition of fibers extracted from waste provide further enhancement in properties. Sustainable practices are the need of the hour, wherein converting a waste material into usable product with enhanced properties is the necessity [13]. Bayerische Motoren Werke AG, trading as BMW Group and Volkswagen AG known internationally as the Volkswagen Group (VW) is using recycled and bio-based composites in their cars, so the market for recycled FRP C is growing [14]. These initiatives could boost car sector profitability, making composite recycling easier. Furthermore, waste generation will rise as demand for glass fiber-reinforced polymer (GFRP) composites grows to 9.18% by 2025. Therefore, recycling solutions are becoming more important to the market, especially for higher price products recycled carbon fibers cost 10 to 20 €/kg compared to 30 to 40 €/kg for virgin carbon fibers, potentially reducing fiber reinforcement costs by 40%. Similarly, the use of recycled carbon fibers in hybrid materials may reduce expenses by up to 70%, fostering a more sustainable and economically viable composite sector. The diminishing lifetime of carbon fiber reinfocd composites, which varies from 2 to 40 years, necessitates the urgent development of advanced recycling systems [15].

The increasing market demand for FRPCs, coupled with heightened awareness of sustainability and circular economy principles, highlights the necessity for ongoing improvements in composite recycling processes and governmental policies to ensure that the industry's growth aligns with environmental objectives. Recycling of composites have a strong impact on sustaianbility. It plays a crucial role to push towards a more sustaianble future. It helps to reduce the waste of composites that ends up in landfill especially in case of composite materails such as glassfiber, carbon fiber abd reinforced palstics which take hundered of years to degrade. Demand for virgin resources are reduced by using recycled materials in the production of new composites. Recycling normally needs less energy than producing a new one which leads to conserve resources reducing carbon footprints. So by recycling reduction of waste is acheived. Recycling of composites helps in material recovery enabling its integration into new composites especially in case of expensive materials like carbon and fiber glass. Due to environmental concern increasing demand for recycling is increased in industries like automative, aserspace etc. Overall, recycling supports a circular economy model where materails are reused /regenerated than disposed off. This cycle of use, recycling, and reuse helps conserve resources and minimizes waste.

4 Overview of Recycled Fiber Sources

From composite point of view recycling can be divided into two subdivisions. One is using recycled fibers for the developemnt of composites and the other is recycling the composites to reuse or refubrish it again. At present the world is facing a challange regarding the disposal of textile waste which is growing fast due to increased poupulation growth, improved living conditions and fast fashion. One of the ways to reduce the massive waste problem can be recycling. Less than 12% of materials in clothing are recycled with 73% -87% either subjected to incinerated or ending up in landfill [16]. This contributes to concerns related to the environment such as air and water pollution and solid waste. These concerns are ever increasing on a daily basis due to the increased production of clothing. Trending fashions changed the rule of demand and supply, leading to increasing clothing productions which was doubled during 2000 and 2015 while consumption decreased by 36%, as well as the projected surge by 63% by 2030, which will lead to an increase in greenhouse gas emissions by 2.7 billion tons annually, highlighting the urgent need for sustainable waste management solutions. The European Union mandates that by 2025, member states must have separate textile waste collection systems to address this issue [17]. So, there is need to utilize theses waste in a sustainable way. A brilliant and workable opening is arising to redesign the life cycle of textile materials by adopting smart practices, selecting smart materials which would make the product more sustainable as well as affordable for the consumer.

Over the last decade or so, recycled textile materials have been considered for reinforcement of composites and additives. It is very important for the environment, industry and cost-effective product manufacturing that the recycled fibers and textile materials replace traditional components of composites. Recycled fiber includes all types of waste that can be post-consumer waste (old clothing etc. discarded by consumers) pre-consumer waste (defected cutting etc. during the manufacturing provcess) and industrial waste (overproduction, obsolete materails of manufacturing). Origion of theses fibers can be natural or synthetic. Recycled fibers generally retain some of the properties of the original material, but they may also exhibit slight degradation in strength, flexibility, and durability, depending on the recycling process used. As per literature, composites made with waste fibers of cotton, are well known for load bearing structure usage, industries like furniture, automobiles etc. and very good mechanical properties [18]. Textile-reinforced cementitious composites (TRCC) are increasingly used in several fields of building technology (e.g. façade panels, roofing, raised floors, and masonry structures) as building elements. One of the promising types of application of waste fiber for these composite materials is promising. The use of recycled materials and waste as cement reinforcement within the building sector can play a relevant role in sustainability, from the environmental, economic, and social perspectives. Extensive research is reported in this area [19].

Composite waste can be used in different applications for composite manufacturing. Also, recently composites materials after their service life are repurposed to create new composite materials, especially in the case carbon fiber composites which are commonly used in automative and aerospace industries. Recycled composites waste can be utilized as fillers in different applications.

5 Recycling Methods

Recycling of waste is the major issue encountered in the way of sustainability/ sustainable products. Collection and processing are the steps involved in the recycling of the composite wastes as most of the wastages are sent to landfill. Recycling composite materials is a developing field that aims to address sustainability challenges [20]. Some methods involved in recycling of composites are Thermal, mechanical and chemical recycling respectively.

- Mechanical recycling: it consists of techniques which involve shredding, crushing the composites into particles or milling it into fine particles.
- Chemical recycling: It consists of different techniques of breaking down the resin chemically to retrieve the fibers. Solvolysis is one of the techniques of the chemical method that involves dissolving the resins from a composite part to reclaim the fibers. It can be used for both thermoplastic and thermoset based composites
- Thermal recycling: In these methods breakdown of composites is done by thermal means and can be used for both types of resins. Pyrolysis is one of the most used methods that involves burning off the resins from a composite part to reclaim the fibers.

High performance composites are those which are designed for high end applications. These high-performance composites constitute high strength/performace polymers and fibers such as carbon, CNTs, Graphene etc. High performance fibers like carbon fiber composites are recycled by pyrolysis which allows for the extraction and reuse of carbon fibers from waste, enabling its integration into new composites. For fiber glass recycling, mechanical and chemical recycling technologies are used. Research is being carried out to find out an alternative possible application of mechanically recycled FRPC. There is an example of glass fiber usage in Portland cement concrete which resulted in significant improvement of its properties with added benefits for the environment. However, in mechanical recycling, the recycled FRP can only be used in downgrade applications as filler or reinforcing agents. There are disadvantages like limited level of benefit of reinforcement as the maximum volume can go up to 10%, and going beyond this the mechanical property is degraded and processing becomes extremely difficult. There are issues of health hazards and safety issues with mechanical recycling as the powder is in large amount along with tiny fiber particles which are produced during grinding or shredding or milling processes. Different recycling methods used for composites also had impact on the environment. Different recycling methods have different issues as in mechnical recycling due to shredding of composite materials it often results in lower quality products make its usage limited for high end applications. Similarly thechemical recycling process often has scalability and cost issues. In thermal recycling methods like pyrolysis the materials ‘ quality is also degraded. Sometimes some processes produce harmful gasses which limit their usage for environemnt sustainability. All the methods are in the research and development phase. Mechanical recycling is the process where waste FRP materials are reused in new composites having better environmental and economic advantages compared to the thermal recycling process. It is reported that among different types of recycling techniques, thermal recycling is the best and feasible option for recycling of carbon and glass fibers-based polymer composites. Efficient use of resources and energy is the main factor to use recycled material. In order to preserve the energy cutting down the energy expenditure invovled in manufacturing of virgin fibers. Replacing virgin fibers with recycled fibers will also result in significant cost reduction. Carbon fibers are used on a very large scale in manufacturing fiber reinforced composites. Virgin Carbon fibers are produced by the high temperature processing and carbonizing of pitch or rayon. This involves a significant amount of energy consumption. It was reported that very high power of approximately 290 Mj/Kg is used to produce virgin carbon fibers. The high demand and escalated use of carbon fibers also produce very huge quantity waste. There is a notable rise of 12% per annum that has been observed in the demand of carbon fibres since 2010. It was predicted that by 2030 the need of carbon fibers will outstrip the global supply of carbon fibers. So, this impairs to utilize the recycled carbon fibers at the end of the life of carbon products. This will definitely reduce the cost as well as environmental impact of carbon fibers compositesClick or tap here to enter text.

6 Challanges During Recycling of Composites

Several challanges occured during recycling of composites due to the materials ‘ complex structure.

Hetrogeneity: Due to variation in materials ‘ properties of matrix(usually a resin) and reinforcement (woven, non woven textiles) the seperation of resin and fiber present a uinique challenge in recycling and other stages of their lifecycle. Also use of diverse materials for reinforcment and resin also create complexity as each type requires different treatment methods. Also resin type leads to a challanging issue. Thermoset resins are normally required breaking down their structure which is difficult and costly.

Strength and Durability: Due to higher strength and durability of composites which is the requirement of the composites, it becomes difficult to recycle.

Fibers damage during recycling: Quality of the textile fiber during recycling is one of the main challenge. This reduces their perforamnce in new composites products.

Econmic viability: Due to high processing costs and the limited market for recycled fibers and matrix materials cost vs recycling value is another issue.

Environmental issues: Some recycling methods also used some chemicals which are not environmentally friendly. Additionally, some processes (thermal recycling) produce harmful emissions and toxic byproducts.

7 Emerging Solutions

Recycling oriented designs: Several approaches are being explored to improve the recycling of reinforced composites, the most important is the development of composites with recycling in mind using such fiber-matrix systems that could be easier to separate. Emphasis is on the development of thermoplastic resins instead of thermoset which could lead to recycling composites easier. Recycling of conventional reinforced composites is always problematic at the end of their life. Recycling processes have become difficult to execute due to potential hazards for the environment as well as legal issues. The solution to the issues related to recycling was answered by use of green matrix based renewable resource or green composite which are generally formulated by embedding the fibers into a bed of bio-polymers that can integrate more readily into natural biodegradation cycles, for example by CO_2-neutral incineration, which includes recovery of energy, or by composting. Emphasis on green composites using natural fibers with biodegradable resins could offer a more environmentally friendly approach.

Advanced recycling technologies: New recycling methods' exploration could help to break down composites more efficiently such as in the chemical recycling recovery method both matrix and fibers are separated but in a recent research a novel microwave-assisted pyrolysis process that can recycle carbon fiber composites more efficiently and cost less than the conventional method [11]. Research into new method like biological recycling methods, such as enzymatic degradation can help a more environmental friendly approach. A Kenyan mealworm larva has been discovered capable of digesting polystyrene, a common plastic pollutant [21]. For carbon and glass fiber composites High Voltage Fragmentation (HVF) is an innovative technique that utilizes high-voltage electrical pulses to selectively break down composite materials. These innovative methods contribute to the sustainable manufacturing of composite fibers by effectively recycling materials and reducing environmental impact.

8 Manufacturing Techniques of Recyled Composites

The process of producing anything/any good using the recycled waste is termed as the remanufacturing technique/method. For the remanufacturing of composites using recycled fibers, fibers waste can be collected in various forms like dry fabric waste, uncured prepregs waste, trimmed fibers, etc. from different sources such as manufacturing waste, end of life waste and extra material. Several conventional and advanced manufacture technologies are used to remanufacture composites from waste materials. Wet lay, compression moulding, resin infusion, vacuum pre-preg and pultrusion are the common methods used for the production of fiber polymers composites. Advanced manufacturing technology like additive manufacturing has a great potential and capability to design composites using the recovered material. So in this section, commonly adopted techniques for remanufacturing of FRPC are briefly discussed/elaborated.

Wet lay up is a facile method to manufacture composites whereby a mold is prepared and according to the desired design the of product the mold is treated to avoid the sticking of fibers or resin with the Mold surface. Then various laminates of fibers are laid, and the resin is applied and rolling is done to squeeze extra resin and then curing is done at room temperature. This method is very suitable to produce composites from recycled or waste fibers. Recycled fibers can be use in any form either fiber mats or small size chopped fibers. it is a very cost effective and traditional method for composite manufacturing [22]. *Hürol Koçoğlu et a.,* produced hybrid composites by the wet lay method, in which scrap carbon fibers were used. It was found that the scrap carbon fibers result in an increase of 60% in the overall flexural strength of composites [23]. Recycled fibers whose length ranges from 1 to 10 mm can be used in the form of nonwoven matts in the wet lay up method. The composite product produced using the wet lay method and recycled carbon fibers can be used in the automotive industry.

Resin infusion is another sophisticated method to produced composites of with better properties at low cost [24]. Another method of composite manufacturing is the *Pultrusion* process where fibers like carbon and glass are pultruded in a thermoset resin bath and then configured into the desired shape using specific shaped dies. The inherent dimensional stability in pultruded composites also gives an extra margin to impart stability into composites produce using the recycled material.

In the Injection Moulding method the fibers are mixed with the thermoplastic resin and then heated to turn it into a viscous liquid then this combined mixture is injected into the mold to form the composite product. This method is suitable to produced\ recycled FRPC as it allows the use of a wide range of fibers and resin. The feasibility of the injection moulding process for development of composites using recycled carbon fibers is also recommended. Injection moulding process allows the usage of short, recycled fibers. The composites produced by the the injection moulding process and recycled fibers exhibit fair properties which are comparable with the virgin fiber [25]. It is note worthy that the recycled fibers whose length is less than 1mm can be used as fillers in composites production by the injection moulding methodClick or tap here to enter text.

Additive manufacturing is an advanced manufacturing technique in which the materials are deposited layer by layer to create a new product. Additive manufacturing method has the potential to create polymer composites from recycled fibers. *3D printing technology and Fused Deposition Modelling (FDM) Printing / printers* also come under its umbrella. 3D printing is a more feasible and consumer friendly manufacturing approach. *Xiaoyong Tian* et al. reported a research work in which 3D printed continuous carbon fibers reinforced PLA composites were recycled and then reused for the remanufacturing of recycled composites [26] FDM technique offers the opportunity to increase the reuse value of carbon fibers. Using recycled carbon fibers in FDM printing along with optimized process parameters tends to increase the mechanical strength of ordinary FDM products [27].

it is well known that recycled fibers are less in length and don't have uniform length distribution so they are subjected to further treatment to properly use them for remanufacturing. For instance, a research work reported in the literature regarding the reuse of recycled fibers for composite manufacturing using FDM technique. rCF obtained in short length from the recycling method were converted into a powder form. This powder form rCF is then mixed with resin pellets and blended in a good way using twin scree extruder to produce filaments. These filaments comprising the blend of recycled fibers and resin filaments are then used to produce composites using an FDM printer. This remanufacturing is very encouraging because it tends to minimize the carbon fibers waste produce from fibers reinforced composites [28]

Performance Evaluation

Performance of composites materials is depicted/influenced by the reinforcement distribution and interfacial bonding between the fibers and polymer matrix. The performance of composites is evaluated by the strength to weight ratio and their durability also subjected under various load conditions. As a general perception the recycled material lacks the properties which is obvious but there are several factors involved in deciding the performance of FRPC. Like Performance of FRPC is significantly influenced by the handling manners adopted during composite manufacturing stage. For example, it was found that the air entrapped at the bonding interface between carbon fibers and epoxy resin is due to inefficient rolling & squeezing. This results in lowering the density of composites due to induced porosity, which ultimately reduced the final physical and mechanical properties. Another important factor is the presence of various types of contaminations in the recycled fibers such as coatings, binders, or fillers and also effect the overall performance of the composites produced using recycled fibers. It is also worth mentioning that sometimes in case of fibers reinforced composites, inverse trend can also be observed. For example, It is reported in literature that the remanufactured fibers composites using recycled CF exhibited 25% better bending strength than the virgin developed composites [29].

Composites produced by recycled CF show lower mechanical properties than virgin carbon fibers. this is because of lower bonding between the fibers and matrix interface. To overcome these issues various surface treatments of recycled fibers are done to enhance the bonding between the fiber and matrix interface. A study reported by *H. cheng* et al., in which the tensile properties of monofilament recycled carbon fibers were evaluated and an increase of 8% tensile strength was observed than the original carbon fibers [27].

9 Life Cycle Assessment

A life cycle assessment (LCA) approach was implemented to evaluate the impact of manufacturing processes made from recycled and natural fibers, separately, over the environment. The analysis was done in order to check, calculate and compare the impact caused by the composites on the environment. In LCA the entire life cycle of a product is considered from the extraction of tje material, manufacturing phase and end of life. The assessment spanned the entire product life cycle from cradle to gate, focusing on contributions to global warming, ozone depletion, ecotoxicity, and resource consumption and emissions associated. In case of FRPC the escalated use of synthetic fibers like carbon, aramid & glass fibers in composite manufacturing pose serious environmental concerns. So, LCA provide a clear and calculated view depicting the impact of FRPC on the environment. For example the life cycle analysis of carbon fibers demonstrates that production of carbon fibers is more energy intensive process/phenomena as it consumes 14 times more energy than the production of conventional steel. Moreover, converting lignin into carbon fiber results in reduction of 20% energy consumption while the direct use of lignin in manufacturing of composite automotive part requires no energy consumption [30]. In another study comparisions among flax fabric (FF), flax–recycled polyamide fabric (F/RPA), and flax–recycled polyester fabric (F/RPES) laminates were done. A comprehensive evaluation covering various stages of their life cycle stages including sourcing of materials which include fiber cultivation and the raw material for bio epoxy resin, processing process of the bio resin and flax fabric, mode and medium of transporatition and process involved for the fabrication process. The assessment was structured around a defined functional unit, which serves as the standard measure for comparing the environmental performance of the composite laminates.

In another study LCA was performed to asses the environmental impact of recycled composite manufacturing methods. Comparison was made to assess the environmental effects of using black recycled ABS pellets in injection moulding and fused deposition modeling (FDM). On employing LCA and Unit process life cycle inventory (UPLCI) methodologies, the analysis revealed that FDM had a lower environmental impact for batch sizes smaller than 14 parts. However, for batches exceeding 50 parts, injection moulding demonstrated lower global warming potential

(GWP) and cumulative energy demand (CED). Mechanical tests on FDM parts with varying infills showed that 100% infill achieved the best balance between mechanical properties and environmental impacts [31].

LCA of products results in thinking about the possible solutions to minimize the effect of production on the ecosystem. The most promising and sustainable solution is replacing the synthetic fibers with natural fibers in composites manufacturing. Using natural fibers in composite manufacturing provided potential solutions in terms of cost, processing and ecofriendly in nature. The other potential solution is use of recycled synthetic fibers like recovered carbon cycles. In any case the energy required for remanufacturing of composites using recycled carbon fibers is less than that of production of polymer composites using virgin carbon fibers. LCA is very important in decision making and understanding the importance of recycling technologies by analysing their positive and negative impacts. For example in chemical recycling of FRPC the use of different solvents, their recycling efficiency, production cost, impact on human health and environment is analysed. In LCA studies it is mentioned that the glass fibers wind turbines ' average life span is 20–25 years but again recovering glass fibers from such waste and reusing them in various applications by retaining their mechanical properties further complete their life cycle loop [32].

10 Case Studies

Due to environmental issues awareness, utilization of finite resource and management of waste disposal made industries to think in a different way on how to use recycled materials. It is crucial to modify composite waste into a valuable resource and close the loop within the composite life cycle. Recycled FRPC are used in various industries which helps to protect the environment and economic viability. Wind turbines blades are reinforced with recycled carbon fiber [33]. Automotive industry BMU group used recycled carbon fiber for the manufacturing of C-pillar and Toyota uses for the manufacturing of the hatch door frame [34]. CarboNXT is a recycled carbon fiber product from Mitsubishi Chemical Advanced Materials (MCAM) used for autoamtive industry. They claimed reduction of Co2 emission and provided cost saving 20–40%over virgin carbon fiber [35]. In another area MCAM is working with Boeing to evalaute thermoplastic composites for aircraft cabin sidewall panels [36]. Composite waste and textile waste are used for the construction industry for various applications [37]. Similarly textile/ garmnet waste are used for different applications like packing materials etc. [38, 39]. These case studies elaborate the diverse applications and utilization of recycled FRPC into various industries highlighting the potenial for waste material usage.

11 Conclusion

Fiber-Reinforced Polymer (FRPC) composites have played an important role in the replacement of metals in applications based on the ease of manufacturing and superiority in quality. There is a diversified trend of its usage primarily due to durability, advanced fabrication technique, easier access of raw material, microstructure and flexibility in design. The recycling of composites and fibers play a pivotal role in supporting the reduction of environmental impact, support of sustainable manufacturing and conservation of resources. The process of recycling is always helpful in the reduction of landfills. It reduces the need for virgin materials, lowering carbon emission associated with the production of materials. As the usage of polymer composites grew, it raised concerns related to waste management. An associated challenge in the use of synthetic polymeric materials in FRPC makes us face recycling and management of waste challenges.

The composites which are conventionally reinforced, whenever it reaches the end of its service life, there is always a serious threat on how to recycle. Due to environmental and legal issues, recycling process has become an increasingly difficult issue. Recycling fibers and composites plays a crucial role in reducing environmental impact, conserving resources, and supporting sustainable manufacturing practices. Recycled fibers, such as PET, fiberglass, and carbon fiber, offer valuable properties, though there are challenges in maintaining their performance during recycling. Most of the high-performance FRPC are based on thermoset polymeric materials, which are non-recyclable. Therefore, fundamental research has been initiated on recycling of thermoset-based FRPC. Research in this area has provided insight into the chemistry of raw materials and techniques of their synthesis and extraction, fabrication, interface chemistry, structural analysis, and microstructural characterizations of FRP composites. It also focusses on the recent progress of FRPC, an alternative to metals for various applications and the challenges faced. The increasing market demand for FRPCs, coupled with heightened awareness of sustainability and circular economy principles, highlights the necessity for ongoing improvements in composite recycling processes and governmental policies to ensure that the industry's growth aligns with environmental objectives. This scnerio overall promotes a more circualr economy approach.

References

1. Jamshaid H, Mishra R (2016) A green material from rock: basalt fiber—a review. J Textile Inst 107(7):923–937.https://doi.org/10.1080/00405000.2015.1071940
2. The global composites market size, 2010–2023. https://www.marketsandmarkets.com/Market-Reports/composite-market-200051282.html#:~:text=. Acessed November 2024
3. Textile Composite Market (2024). https://www.mordorintelligence.com/industry-reports/textile-composites-market

4. Amaechi C, Odijie C, Sotayo A, Job S, Wang F, Hou X (2019) Recycling of renewable composite materials in the offshore industry. https://doi.org/10.1016/B978-0-12-803581-8.11445-6. In book: reference module in materials science and materials engineering
5. Composites Market: Trends, Opportunities and Competitive Analysis [2024–2030] Global composites market report 2019: $40.2 billion market/ composites market: trends, opportunities and competitive analysis [2024–2030]
6. Naqvi SR et al (2018) A critical review on recycling of end-of-life carbon fiber/glass fibre reinforced composites waste using pyrolysis towards a circular economy. Resour Conserv Recycling 136:118–129
7. European Composites Industry Association (2017) Composites and sustainability—the big picture, pp 1–5, 2017, [Online]. Available: https://eucia.eu/userfiles/files/EuCIAblog_Sustainability_the big picture_November 2017.pdf
8. Aerospace Composites Market Size, Share, Trends, Industry Analysis Report—By Fiber Type (Glass Fiber, Carbon Fiber, Ceramic Fiber, and Other Fiber Types), By Matrix Type, By Manufacturing Process, By Aircraft Type, and By Region; Segment Forecast, 2024–2032, Aerospace Composites Market Size, Share, Growth | Report, 2032
9. Sudhin AU, Remanan M, Ajeesh G, Jayanarayanan K (2020) Comparison of properties of carbon fiber reinforced thermoplastic and thermosetting composites for aerospace applications. Mater Today Proc 24(Part 2):453-462, ISSN: 2214-7853. https://doi.org/10.1016/j.matpr.2020.04.297
10. Mäkelä M, Rissanen M, Sixta H (2020) Machine vision estimates the polyester content in recyclable waste textiles. Resour Conserv Recycling 161:105007, ISSN: 0921-3449.https://doi.org/10.1016/j.resconrec.2020.105007
11. Serra A, Tarrés Q, Llop M, Reixach R, Mutjé P, Espinach FX (2019) Recycling dyed cotton textile byproduct fibers as polypropylene reinforcement. Text Res J 89:2113–2125
12. Tufail MR, Jamshaid H, Mishra R, Hussain U, Tichy M, Muller M (2021) Characterization of hybrid composites with polyester waste fibers, olive root fibers and coir pith micro-particles using mixture design analysis for structural applications. Polymers 13(14):2291. https://doi.org/10.3390/polym13142291
13. Amaechi C, Odijie C, Orok E, Job S (2019) Economic aspects of fiber reinforced polymer composite recycling. In book: Reference module in materials science and materials engineering. https://doi.org/10.1016/B978-0-12-803581-8.10738-6
14. Zabihi O et al (2020) A sustainable approach to the low-cost recycling of waste glass fibres composites towards circular economy. Sustainability 12(2):641
15. Thomason JL, Nagel U, Yang L, Sáez E (2016) Regenerating the strength of thermally recycled glass fibres using hot sodium hydroxide. Compos Part A Appl Sci Manuf 87:220–227, ISSN: 1359-835X. https://doi.org/10.1016/j.compositesa.2016.05.003
16. Clark JH (2023) Textile waste—an opportunity as well as a threat. Green Carbon 1(2):146–149. ISS:N 2950-1555. https://doi.org/10.1016/j.greenca.2023.10.002
17. Overview of all Textile Extended Producer Responsibility (EPR) Laws (2025) Overview of all textile extended producer responsibility (EPR) Laws. Accessed January 2025
18. Petrucci R, Nisini E, Puglia D, Sarasini F, Rallini M, Santulli C, Minak G, Kenny J (2015) Tensile and fatigue characterisation of textile cotton waste/polypropylene laminates. Compos Part B Eng 81:84–90
19. Jamshaid H, Shah A, Shoaib M, Mishra RK (2024) Recycled-textile-waste-based sustainable bricks: a mechanical, thermal, and qualitative life cycle overview. Sustainability 16:4036.https://doi.org/10.3390/su16104036
20. Current Trends in the Use of Recycled Carbon Fiber (2024) Current trends in the use of recycled carbon fiber. Accessed on December 2024
21. Plastic-eating worm discovered— could help reduce pollution 'faster and more efficiently', scientists say (2024) Plastic-eating worm discovered that could help reduce pollution, scientists say. Acessed on December 2024
22. Mourad AH, Zaaroura N, Cherupurakal N (2019) Wet lay-up technique for manufacturing of advanced laminated composites. Karbala Int J Modern Sci 5(1). https://doi.org/10.33640/2405-609X.1098

23. Koçoğlu H, Kodal M, Altan MC, Özçelik B, Özkoç G (2022) A new approach for the reuse of scrap carbon fiber in high-added value continuous fiber reinforced composite structures. Compos Part A Appl Sci Manuf 163:107272. https://doi.org/10.1016/j.compositesa.2022.107272
24. Gajjar T, Shah DB, Joshi SJ, Patel KM (2020) Analysis of process parameters for composites manufacturing using vacuum infusion process. Mater Today Process 21:1244–1249. https://doi.org/10.1016/j.matpr.2020.01.112
25. Rimmel O, May D, Goergen C, Poeppel A, Mitschang P (2019) Development and validation of recycled carbon fiber-based binder tapes for automated tape laying processes. J Compos Mater 53(23):3257–3268. https://doi.org/10.1177/0021998318820422
26. Tian X, Liu T, Wang Q, Dilmurat A, Li D, Ziegmann G (2017) Recycling and remanufacturing of 3D printed continuous carbon fiber reinforced PLA composites. J Clean Prod 142:1609–1618. https://doi.org/10.1016/j.jclepro.2016.11.139
27. Cheng H, Guo L, Qian Z, Sun R, Zhang J (2021) Remanufacturing of recycled carbon fiber-reinforced composites based on fused deposition modeling processes. Int J Adv Manuf Technol 116(5–6):1609–1619. https://doi.org/10.1007/s00170-021-07473-w
28. Ateeq M (2023) A state of art review on recycling and remanufacturing of the carbon fiber from carbon fiber polymer composite. Compos Part C Open Access 12. https://doi.org/10.1016/j.jcomc.2023.100412
29. Das D, Dubey OP, Sharma M, Nayak RK, Samal C (2019) Mechanical properties and abrasion behaviour of glass fiber reinforced polymer composites—a case study. Mater Today Proc 19:506–511. https://doi.org/10.1016/j.matpr.2019.07.644
30. Das S (2011) Life cycle assessment of carbon fiber-reinforced polymer composites. Int J Life Cycle Assess 16(3):268–282. https://doi.org/10.1007/s11367-011-0264-z
31. Al Rashid A, Khan SA, Koç M (2024) Life cycle assessment on fabrication and characterization techniques for additively manufactured polymers and polymer composites. Cleaner Environ Syst 12. https://doi.org/10.1016/j.cesys.2023.100159
32. Karuppannan Gopalraj S, Kärki T (2020) A review on the recycling of waste carbon fibre/glass fibre-reinforced composites: fibre recovery, properties and life-cycle analysis. Springer Nature.https://doi.org/10.1007/s42452-020-2195-4
33. Chen J, Wang J, Ni A (2019) Recycling and reuse of composite materials for wind turbine blades: an overview. J Reinforced Plastics Compos 38:567–577. https://doi.org/10.1177/0731684419833470
34. Utekar S, Suriya VK, More N, Rao A (2020) Comprehensive study of recycling of thermosetting polymer composites—driving force, challenges and methods. Compos Part B Eng 207:108596. https://doi.org/10.1016/j.compositesb.2020.108596
35. Bledzki AK, Seidlitz H, Krenz J, Goracy K, Urbaniak M, Rösch JJ (2020) Recycling of carbon fiber reinforced composite polymers—Review—Part 2: Recovery and application of recycled carbon fibers. Polymers 12:3003. https://doi.org/10.3390/polym12123003
36. CarboNxt$^{(R)}$. 100% CUSTOMISED 100% PURE 100% RECYCLED
37. Design flexibility and performance meet in KyronTEX®. KyronTEX®—Drapable, Recyclable Carbon Fiber Composite | MCG
38. Ali H, Jamshaid H, Mishra R, Chandan V, Jirku P, Kolar V, Muller M, Nazari S, Shahzada K (2023) Optimization of seismic performance in waste fibre reinforced concrete by TOPSIS method. Scientific Reports 21;13(1):8204. https://doi.org/10.1038/s41598-023-35495-9. Erratum in: Sci Rep. 2023 Jun 20;13(1):10018. https://doi.org/10.1038/s41598-023-35495-9. PMID: 37211550; PMCID: PMC10200795
39. Jamshaid H, Mishra R, Zeeshan M, Zahid B, Basra SA, Tichy M, Muller M (2021) Mechanical performance of knitted hollow composites from recycled cotton and glass fibers for packaging applications. Polymers 13:2381. https://doi.org/10.3390/polym13142381

Recycling of Natural Fibre-Reinforced Polymer Composites (NFRPCs): Challenges and Opportunities

Muhammad Kashif Bangash and Maurice N. Collins

Abstract A composite material is fundamentally composed of at least two distinct materials. This definition includes natural substances like bricks, concrete, wood, and bone, as well as engineered composites such as fibre-reinforced plastics (FRP).

1 Introduction

A composite material is fundamentally composed of at least two distinct materials. This definition includes natural substances like bricks, concrete, wood, and bone, as well as engineered composites such as fibre-reinforced plastics (FRP). FRPs are widely used in various industries for structural applications due to their high strength, stiffness, and low weight. FRPs, such as those made with high performance fibres like aramid, carbon, and glass fibres, are widely used in industries like automotive, construction, and aerospace due to their superior mechanical properties and lightweight. However, synthetic fibres pose environmental risks and are non-biodegradable. To address these concerns, natural fibres are becoming more popular as eco-friendly alternatives in composite materials. Natural fibres are sourced from plants (linen, hemp, jute, cotton), animals (wool, silk), and minerals (asbestos) [1].

Natural fibres are available in a variety of types, including bast fibres (flax, jute, ramie, hemp, and kenaf), seed fibres (cotton, coconut, and kapok), and leaf fibres (pineapple, abaca, and sisal). Agricultural residues from grasses and reeds, such as corn, rice, and wheat, are used as filler reinforcements in composites. The health risks associated with certain natural fibres, such as the carcinogenic effects of asbestos, have led to their prohibition in numerous countries. Figure 1 shows the key natural fibres commonly used as reinforcement in composite materials.

M. K. Bangash (✉) · M. N. Collins
School of Engineering, Bernal Institute, University of Limerick, Limerick, Ireland
e-mail: Kashif.bangash@ul.ie

M. N. Collins
e-mail: maurice.collins@ul.ie

H. Jamshaid and U. Zubair (eds.), *Sustainable Recycling of Fiber Reinforced Polymer Composites*, SDGs and Textiles, https://doi.org/10.1007/978-981-96-9536-2_2

Fig. 1 Examples of key natural fibres used as reinforcement in composites

Natural fibres are abundantly available. Table 1 lists the global production, content, and mechanical properties of key natural fibres used in composites. Natural fibres used as reinforcement in FRPs can be classified by their sources. The primary group of natural fibres includes fibres like jute, hemp, kenaf, and sisal, which are cultivated primarily for their fibre content. The secondary group comprises fibres like pineapple, oil palm, and coir, which are typically byproducts of other processes.

Different types of natural fibres have unique properties that make them suitable for specific applications. For instance, bast fibres have good mechanical properties and show excellent bonding with polymer matrix are derived from the outer layers of plant stems through retting, where bacterial activity and moisture help separate

Table 1 Global production, content, and mechanical specification of key natural fibres used in composites [6]

Type of fibre	Production (10^3 tonnes)	Chemical composition (%)			Tensile strength (MPa)	Elongation at break (%)	Young's modulus (GPa)
		Cellulose	Hemi-cellulose	Lignin			
Jute	2,850	61–63	13	5–13	200–800	1.16–8	10–55
Banana	200	60–65	6–8	5–10	529–914	3	27–32
Coir	650	43	<1	45	106–175	14.21–49	4–6
Flax	830	70–72	14	4–5	300–150	1.3–10	24–80
Kenaf	970	70	8	12	295–119	3.5	2.86
Pineapple	–	80	–	12	170–162	2.4	60–82
Sisal	318.8	60–67	10–15	8–12	80–840	2–25	9–38
Hemp	214	70–78	18–19	4–5	310–900	1.6–6	30–70
Ramie	100	80–85	3–4	0.5	348–938	1.2–8	44–128

the fibres. Leaf fibres are known for their high strength and are extracted by scraping leaves either manually or mechanically. Seed fibres are collected from seeds or fruits, are lightweight, soft, and offer excellent insulation. Stem fibres are sourced from plant stalks like sugarcane, corn, and wheat, are thick, tough, and low-density. Grass fibres are obtained from tall grasses like bamboo, are known for their strength, flexibility, and lightness. Tree fibres are derived from softwood and hardwood trees which are often mixed with virgin or recycled fibres to make them adaptable for use in composites and paper production [2–4].

Natural fibres have a significantly lower carbon footprint (0.3–0.5 tonnes CO_2-eq) compared to glass fibres (1.7–2.5 tonnes CO_2-eq) during fibre production. However, during composite manufacturing, natural fibres exhibit a slightly higher carbon footprint than glass fibres, partially offsetting their environmental advantage. Natural plant fibres have been incorporated into FRP composites, termed as biocomposites, for applications such as automotive interiors, construction, insulation, and transport. Efforts to advance sustainable practices are focused on developing lightweight, cost-effective fibres for structural composites using renewable and bio-based materials. These eco-friendly fibres minimize environmental impact through controlled chemical usage, reduced resource consumption, and biodegradability [5].

The matrix in a composite material serves critical functions such as transferring applied stresses to the reinforcing fibres, protecting fibres from environmental and mechanical damage and maintaining the overall structural integrity. The matrix material must show enough deformability and compatibility with the reinforcement to ensure the desired mechanical properties. Strong adhesion between the matrix and fibres is essential for efficient stress transfer and minimizing relative movement. In most FRP composite systems, the matrix is a polymeric resin which consists of large macromolecules formed by the covalent bonding of monomer units. The degree of polymerization, or the number of monomers in a polymer chain, influences its properties. Figure 2 shows the chemical structure of monomers of key polymers used as matrix in FRPs. Natural macromolecules include materials like rubber, polysaccharides, DNA, and proteins, while synthetic macromolecules include polymers such as polyethylene, polypropylene, polystyrene, PVC, PTFE, polyesters, polycarbonates, polysiloxanes, and polyimides. Each type of polymer matrix offers unique characteristics suited to specific applications [7].

Thermosets and thermoplastics are two major categories of polymer resins, with distinct properties and behaviours during processing, used in composites and various applications. Thermoplastics are polymers with linear chains connected by weak bonds which makes them easy to shape by heating and cooling. They are widely used due to their low cost, recyclability, and thermal resistance. These resins are often processed using injection moulding, extrusion, or thermoforming techniques. PVC and polystyrene are common examples of thermoplastic resins. On the other hand, Thermosetting polymers form solid, infusible networks through chemical reactions with a hardener. Epoxy resins and unsaturated polyesters are most common thermoset resins. Compared to thermoplastic composites, thermoset matrix materials offer superior modulus, creep resistance, and thermal and chemical stability. However, they tend to be more brittle and have lower fracture toughness at room temperature. Common thermosets used in fibre-reinforced composites include epoxy, vinyl ester, unsaturated polyester, and phenolic resins [8].

Natural fibre-reinforced polymer composites (NFRPCs) are produced using conventional polymer composite fabrication techniques such as hand lay-up, vacuum resin infusion, resin transfer moulding, compression moulding, injection moulding, additive manufacturing, extrusion, and pultrusion. The processing technologies for shaping composites are similar for both natural and synthetic fibre composites, the major disadvantage of natural fibres in composite fabrication is the maximum processing temperature, as natural fibres begin to degrade at temperatures above 200–230°C. When natural fibres are used as reinforcement and the matrix made of bio-based or biodegradable resins, the material is typically referred to as a biocomposite. In contrast, composites designed to be more environmentally friendly throughout their lifecycle are called green composites. Green composites can incorporate either natural or synthetic fibres but must have an environmentally conscious matrix [9].

Despite their many benefits, NFRPCs face challenges that limits their widespread use in advanced applications. One major challenge is the presence of predominantly hydrophilic cellulose in natural fibres which contrasts with the hydrophobic properties of most polymers. This mismatch affects the interfacial bonding between the fibres and the polymer matrix and degrade the mechanical properties of the composite; however, this effect can be reduced by chemical treatment of natural fibres. Another issue of natural fibres is lower mechanical strength compared to synthetic counterparts. Natural fibres such as flax, hemp, jute, sisal, and coir display significantly high strength and are viable alternatives to glass fibres in moderate-strength applications. These fibres offer additional benefits like high specific strength, thermal insulation, acoustic absorption, and biodegradability, which contribute to reducing the environmental impact of composite materials. However, these fibres still fall short in performance when compared to high-performance synthetic fibres

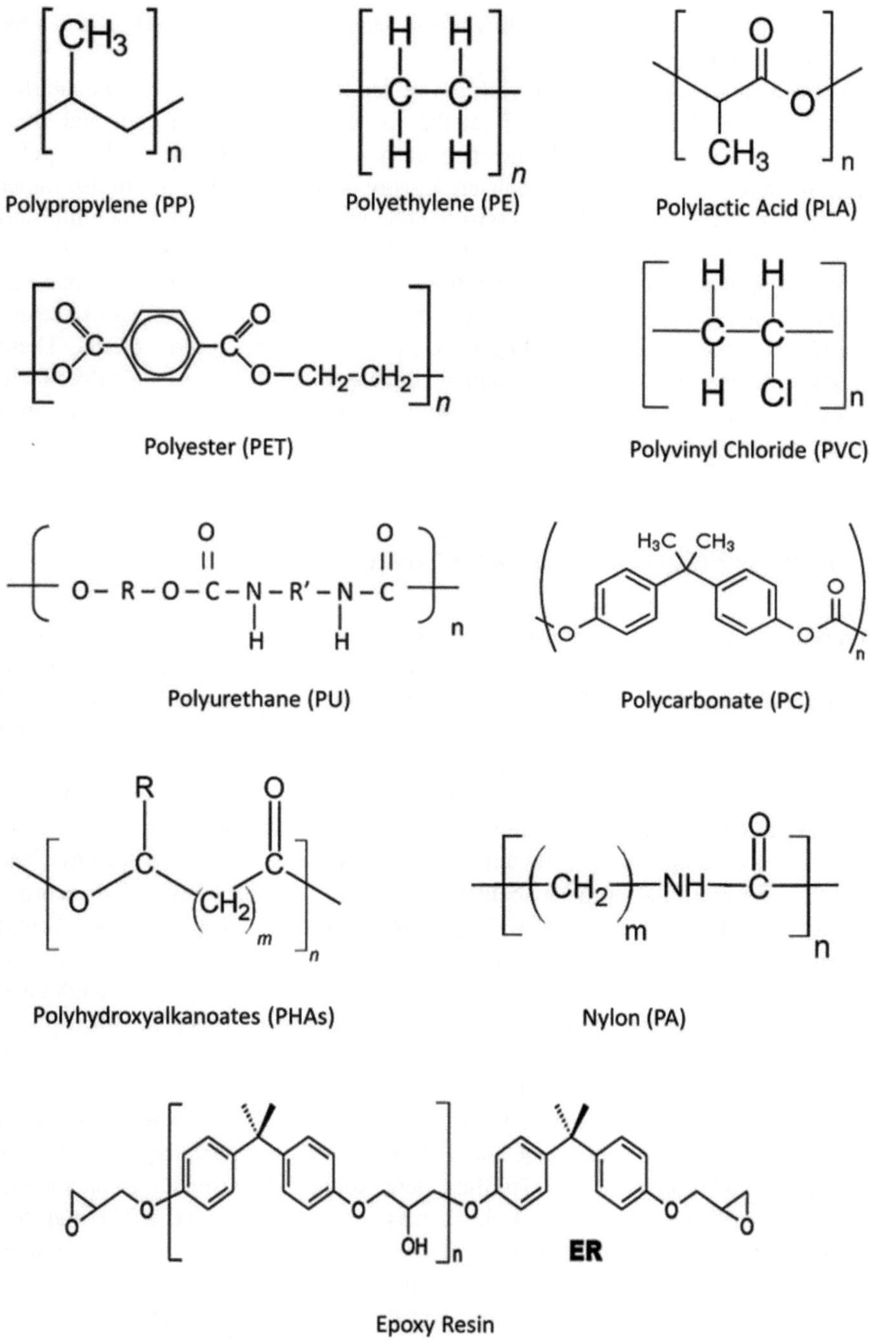

Fig. 2 Chemical structure of key polymers used as matrix in composite materials

such as Nomex for fire resistance, Kevlar for shear resistance, and carbon fibres for high specific strength [5].

Though natural composite materials are valued for their durability, sustainability, lightweight, and high performance yet they pose significant environmental challenges due to the non-recyclable and chemically inert nature of their synthetic polymeric matrices. The matrix materials are mostly designed for long-term use rather than biodegradability, resulting in considerable waste accumulation in landfills and oceans. Currently, the widespread adoption of FRP composites across industries has led to a significant rise in waste generation. To combat this, industries are increasingly advocating for a circular economy approach, emphasizing practices like reuse, repair, remanufacture, and recycling through the 'make-use-return' model. There is a pressing need to develop environmentally sustainable and economically viable methods for managing and recycling composite waste to reduce their long-term impact on the environment.

2 Fibre Reinforced Composite Waste

Composite materials as reinforced plastics are widely used with more than 40,000 industrial products in energy, transportation, medical, sports, aerospace, and construction industries. Their demand continues to grow, particularly for industrial applications. The global automotive composites market was valued at USD 7.67 billion in 2018 and is projected to grow to USD 13.5 billion by 2026, reflecting a compound annual growth rate (CAGR) of 7.56% during the forecast period. In 2018, North America held the largest share of the market, accounting for 31.29%. Around 45% of major components in modern airplanes are now made of composites. The Boeing 787 Dreamliner utilizes about 50% while the Airbus A350 XWB incorporates around 53% composite materials contributing to their lightweight and efficient design. However, this increasing usage has raised environmental concerns due to waste management issues, rising dependency on fossil fuel prices, and heightened awareness of ecological impact among consumers and legislators [10, 11].

Over 350 million metric tonnes of plastic waste are produced each year which are expected to triple by 2060 and will reach a staggering one billion metric tonnes if no significant efforts are made. With the introduction of stricter measures, such as increased taxes on plastic usage and a global overhaul of recycling systems, this projection could drop to under 700 million metric tonnes. However, it is a challenging goal as currently, less than 10% of plastic waste is recycled annually. The majority ends up in landfills or incinerators, releasing harmful pollutants into the environment, while about a quarter is either mismanaged or littered [12].

Mismanaged plastic waste is mostly dumped illegally, burned in open pits, or leaks into natural environments, including rivers and oceans. Between 1970 and 2019, an estimated 30 million metric tonnes of plastic accumulated in the ocean, with over 100 million tonnes in rivers and lakes. This level of pollution has devastating effects on marine ecosystems. Globally, underdeveloped regions are responsible for over

80% of global plastic waste discharged into the ocean but the wealthier regions contribute around 5% to the global plastic waste; however, these regions export significant volumes of plastic waste to developing countries, many of which lack adequate infrastructure to handle such large amounts, exacerbating the global waste crisis [13].

3 Recycling of Natural Fibre-Reinforced Polymer Composites

Fibre-reinforced polymer composites (FRPCs) are widely used in industries like automotive, aerospace, and renewable energy due to their strength, lightweight, and durability. The automotive sector, especially with the rising production of EVs and hydrogen-powered vehicles, is expected to see significant growth in the use of composites in coming years. This growing demand of composites highlights the importance of managing waste from these materials in an environmentally friendly way. Figure 3 represents the estimated FRP composite waste produced in 2025. To avoid the undesired plastic waste accumulation in the environment, several techniques and methods are proposed and experimented in the past. Estimates suggest that 80% of global plastic production is comprised of thermoset and thermoplastics resins. Thermoplastics are generally easy to recycle and can be reused. On the contrary, thermosets are difficult to recycle due to their network structure. More than 90% of all FRP products use glass fibres while the remaining 10% use other synthetic or natural fibres, and 66% of all composites use thermoset resin [14].

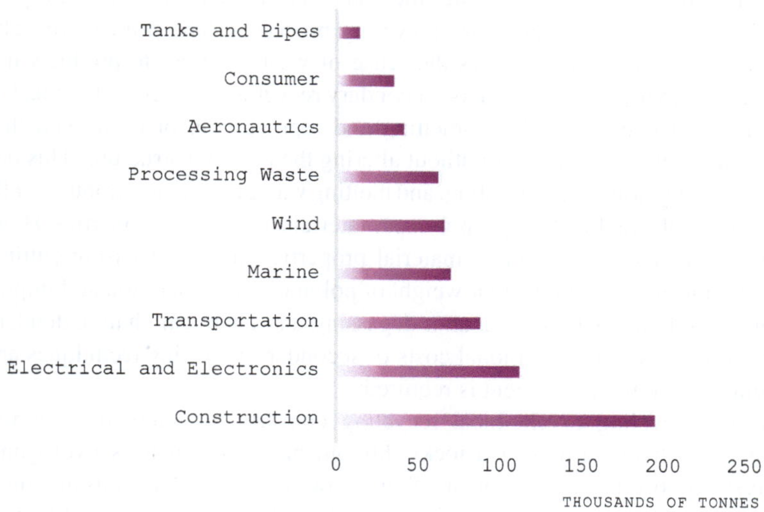

Fig. 3 Estimated FRP composite waste in thousands of tonnes in 2025, adapted from [15]

Traditional methods like landfilling and incineration are harmful to the environment and face restrictions due to regulations like the EU Directive 2018/850, which aims to limit landfill use. To deal with FRPC waste, the EU's waste management hierarchy recommends first minimizing and preventing waste, followed by reuse, recycling, energy recovery, and, as a last resort, disposal [16]. Reusing and repurposing FRP parts at the end of their life cycle can further reduce waste. For example, components like domes and chimneys are easier to reuse, whereas structural profiles may be harder due to deterioration or unknown material properties. Challenges in reuse include identifying the condition of the materials and the high costs of reclaiming fibres and resins. Recycling these materials can be challenging due to cost, recycled material quality, lack of reprocessing facility, and well-established market for recycled products.

Besides the end of life, FRP waste from processing and production can be reduced by using efficient manufacturing methods, such as pultrusion or filament winding, which produce less scrap compared to manual techniques like hand layups. For instance, automated methods typically result in 2–5% waste, while manual processes can create up to 15%.

By improving manufacturing processes, enhancing recycling methods, and creating markets for recycled FRPC materials, we can reduce waste and lessen the environmental impact of these composites. These efforts are crucial for meeting the growing demand for composites while promoting sustainability.

4 Recycling Systems

Polymer and composite recycling methods can be categorized into primary, secondary, tertiary, and quaternary recycling processes. The primary recycling is basic, cost-effective, and involves the reuse of waste composite products in their original form to make new products. Secondary recycling also termed as mechanical recycling, and it involves the re-melting and reprocessing of reinforced thermoplastic waste into new products without altering the polymer structure. This process includes sorting, washing, shredding, and melting waste to produce granules or flakes, which can be blended with virgin polymer materials to enhance properties. However, in this type of recycling reduces material properties due to excessing cutting and chopping, lowering the molecular weight of polymers chain scission and impurities. However, methods including vacuum degassing and chemical chain extenders can ease these effects. The operational costs of secondary recycling techniques are low but significant initial investment is required.

Tertiary recycling or chemical recycling converts polymers into monomers, oligomers, or hydrocarbon feedstocks through processes such as hydrogenation, glycolysis, pyrolysis, and hydrolysis. Some processes, like glycolysis and metanalysis, are commercially implemented, while others, like gasification and microwave pyrolysis, remain under investigation due to high capital and technical expertise

Table 2 Four main types of recycling processes are used in the composites recycling industry

Sr#	Recycling process	Description
1	Mechanical processes	The material is reintegrated into a new object without altering its chemical structure. After shredding, the material is separated into fibre-rich and matrix-rich fractions, which typically have lower quality than the original material
2	Thermal processes	The material is treated at high temperatures, resulting in the recovery of solid residues and energy
3	Chemical processes	Involves chemical depolymerization, where solvents are used to separate the organic and mineral components. These can be used as intermediate products in new syntheses or products. This method has environmental concerns due to potential chemical waste. Supercritical fluids offer a cleaner alternative
4	Biological processes	This method involves the degradation of materials by living organisms. However, fibre-reinforced polymers are designed for long lifespans and are not biodegradable, so this route is not very common

requirements. The quaternary recycling involves energy recovery through the incineration of polymer waste. The techniques are commercially well established but results in harmful emissions and residues [6]. Composites recycling processes are mainly classified into four types: mechanical, thermal, chemical, and biological as discussed in Table 2.

4.1 Mechanical Recycling

Mechanical recycling involves breaking down of NFRPCs waste into smaller particles, about 20–200 mm in size, and then grinding them into fine particles through grinding, shredding, or milling. This process produces two main outputs:

- Fibrous fraction containing natural fibres.
- Fine powder primarily composed of the resin matrix.

The typical required equipment for mechanical recycling includes a basic rotary cutter/granulator, high speed mill, a pulverizer for crushing the ground material in powdered form, and/or a classifier also known as cyclones and sieves, or an electrostatic separator for separating the coarse fibrous and fine powdered products. Figure 4 represents the classification of recycling techniques for FRP composites.

The grinded particles thermoplastic composites reinforced with natural fibres like hemp, flax, jute, sisal, or kenaf are extruded into pallets for reuse. The coarse products contain fibrous material and can be reused in bulk moulding compounds (BMC) sheet moulding compounds (SMC) which are traditionally associated with thermosetting resins such as polyester, vinyl ester, or epoxy and glass fibres. However, recent

Fig. 4 Classification of recycling processes and techniques for FRP composites

advancements have expanded their applicability to thermoplastics and natural fibres. Figure 5 represents the steps of mechanical recycling process of FRP composite.

The fibrous fraction of recycled NFRPs can serve as reinforcement in the production of new composites and the powdered fraction can be reused in applications such as adhesives or low-strength components. Mechanical recycling offers a cost-effective and scalable solution for NFRCs; however, it leads to a reduction in fibre length and polymer quality. This degradation limits the performance and value of recycled materials.

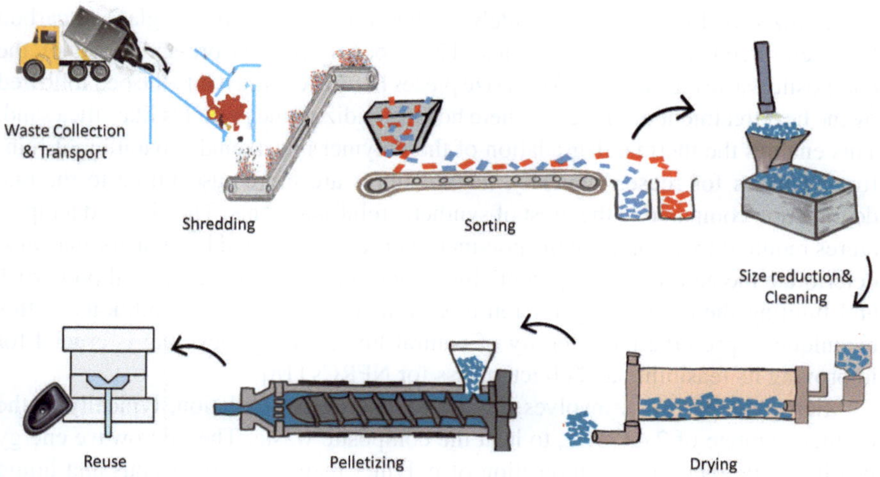

Fig. 5 Mechanical recycling of FRP composites, adapted from [17]

4.2 *Thermal Recycling*

Thermal recycling separates fibres from the resin matrix through heat-based methods such as pyrolysis, fluidized beds, and microwave pyrolysis. Pyrolysis involves heating of FRPs waste in an oxygen-free environment, decomposing the resin into gaseous, oily, and solid byproducts. This method is particularly suited for synthetic fibre-reinforced composites since matrix polymers in composite can serve as an energy source while high value synthetic fibres are effectively reclaimed for reuse in various applications. Figure 6a shows a typical set-up for FRP pyrolysis process.

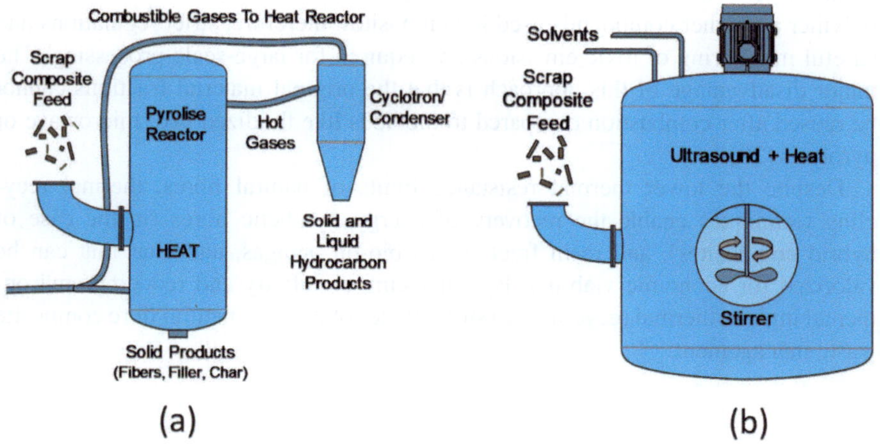

Fig. 6 **a** Pyrolysis, thermal process; **b** dissolution, chemical process. Adapted from [22]

Fluidized bed technology is widely studied for the recycling of glass or carbon fibre-reinforced polymer composites. The process requires pre-shredding of the composite waste into around 25 mm size pieces before feeding it into the bed followed by the heat treatment in a reactor where hot air fluidizes fine particles like silica sand. This enables the thermal degradation of the polymer matrix and separation of reinforcing fibres for reuse. Typically, natural fibres are more susceptible to thermal degradation compared to the most of synthetic reinforcements. The elevated temperatures required to decompose the polymer matrix in a fluidized bed reactor may also deteriorate the natural fibres, potentially compromising their mechanical properties and limiting their suitability for reuse. Optimizing the process conditions of this technique to preserve the integrity of natural fibres during recycling is crucial for improving its feasibility and effectiveness for NFRCs [18].

Microwave pyrolysis involves the use of microwave radiation, typically in the frequency range of 2.45 GHz, to heat the composite waste. The microwave energy facilitates the thermal decomposition of polymer matrices into gaseous and liquid byproducts, while the reinforcing natural fibres are subjected to thermal degradation. Additives such as carbon black or biochar are used to enhance the heating efficiency of the materials having low microwave absorption properties. This technique is ideal for clean fibre recovery and valorization of the matrix fractions. However, exposure to elevated thermal conditions often degrade the mechanical properties of natural fibres more severely than synthetic ones which can be challenging for the recovery of such fibres [19].

Incineration is another thermal process that involves burning composite waste to generate heat energy and is efficient for handling composite waste with complex coatings or paint, which are difficult to recycle conventionally. This method is often used for highly contaminated waste that cannot be recycled through conventional or available means. Incineration can reduce up to 96% of solid waste by weight. However, it is less environmentally friendly compared to mechanical recycling due to the emission of significant amounts of CO_2, and hazardous gasses based on matrix polymer and other compounds used in composites, therefore, strict regulations and careful monitoring of toxic emissions are required for large-scale processing. The major disadvantage of this approach is that the original material fractions cannot be reused after combustion compared to methods like fluidized bed, microwave or pyrolysis [20].

Despite the lower thermal resistance limits of natural fibres, thermal recycling techniques enable the recovery of energy, synthetic fibres (in the case of hybrid composites), and resin fractions as bio-oil, syngas, and char that can be valorized for economic viability. By enhancing circularity and reducing environmental impact, thermal recycling is a sustainable solution for natural fibre composite waste management.

4.3 Chemical Recycling

Chemical recycling of NFRCs is a involves degradation of the polymer matrix into fraction like monomers, oligomers, or other valuable chemicals through various chemical reactions, while aiming to preserve or partially recover the natural fibres for potential reuse. Several chemical recycling approaches have been adopted by researchers in recent past including hydrolysis, glycolysis, methanolysis, and pyrolytic depolymerization. Figure 6b shows a typical set-up for chemical recycling of FRP composites.

Hydrolysis implies breaking down of polymer matrix in the presence of water, sometimes under high temperature and pressure. Hydrolysis is efficient in breaking down the bio-based polymer matrices into organic acids and other monomers, while preserve the natural fibres such as flax or hemp. Alkaline hydrolysis of fibre reinforced PET and PVC composites uses 4–20% NaOH or KOH solutions at temperatures of 120–200 °C (aqueous NaOH) or 110–120°C (non-aqueous KOH). It is crucial to optimize the process conditions to avoid degradation of the fibres [21].

Glycolysis involves the use of glycols such as ethylene glycol to depolymerize thermoset resins, especially those based on unsaturated polyesters or epoxies. Glycolysis has been applied to recover resin fragments and natural fibres, but the exposure to elevated temperatures may cause partial thermal degradation of the fibres.

Methanolysis is another depolymerization technique where methanol acts as a reagent. It is particularly effective for certain bio-based polymers and can yield monomers like dimethyl terephthalate (DMT) and ethylene glycol (EG) [23]. For NFRCs, optimizing methanolysis parameters and conditions to protect fibres remains a critical focus.

Chemical recycling has shown significant potential for the recycling of NFRCs; however, further research is required to optimize these processes particularly to preserve natural fibre properties and upscaling to improve economic feasibility. Recent developments in chemistry and the development of low-energy, selective depolymerization methods hold significant potential to enable large-scale implementation of chemical recycling for NFRCs.

4.4 Biological Recycling

Biological recycling of NFRCs is an emerging method, primarily applicable to composites with biodegradable resins. This process involves the degradation of composite materials by living organisms such as fungi, bacteria, or enzymes. Biodegradation occurs through the action of microbial species and enzymes on the surface of polymer resins. Bacteria and fungi adhere to the FRP surface, release enzymes, and grow by using the polymeric resin as a substrate and nutrient source. This process gradually depolymerizes the polymers leading to mineralization while, water (H_2O), carbon dioxide (CO_2), and methane (CH_4) are produced as final byproducts. Conventional NFRCs with non-biodegradable polymers, such as epoxy or polyester, are less compatible due to their long-term durability. The biodegradation recycling processes may consume or affect the morphological and structural quality of natural fibres [24].

Biodegradation of polymers can occur under both aerobic and anaerobic conditions. In aerobic biodegradation, oxygen serves as the electron acceptor, with aerobic fungi breaking down complex organic compounds into smaller ones, producing CO_2 and H_2O as byproducts. Sturm test is common method to measure CO_2 formation in lab tests to assess polymer degradation. Anaerobic biodegradation occurs without oxygen and the microbes use alternative electron acceptors like sulphate, nitrate, iron, manganese, or CO_2. This process produces CH_4, CO_2, H_2O, and biomass as byproducts. Certain microbes, such as Pestalotiopsis microspora isolates have demonstrated the ability to degrade polyurethane (PUR) effectively under both aerobic and anaerobic conditions, with enhanced activity observed in anaerobic environments [25].

Several biodegradable resin systems have been developed to improve the feasibility of biological recycling for natural fibre-reinforced composites (NFRCs). Composites made from resins derived from renewable sources, such as polylactic acid (PLA), plant oil-based PUR, Epoxidized Soybean Oil (ESO), and bio-based polyesters like PBAT, are gaining attention for this reason. These resins are more responsive to microbial degradation under specific environmental conditions which makes them ideal candidates for biological recycling.

Biological recycling has emerged as a promising strategy for recycling a variety of commonly used polymer matrices. Several microbial species have been identified for their ability to degrade various plastics which can be exploited as a sustainable recycling alternative. Table 3 provides an overview of microbial species and enzymes capable of degrading various polymer matrices. The efficiency of degradation often depends on environmental parameters such as temperature, moisture, pH and the pretreatment of plastics. Degradation rates can also vary between different microbial species.

Despite these advances, the commercial adoption of biological recycling is hindered by slow degradation rates and the need for specific conditions to promote microbial activity. To make this approach more viable on a larger scale, further research is needed in faster-degrading bio-resins and improved biological agents. The biodegradability of polymers in composites is influenced by their chemical structure, molecular weight, and the environmental conditions for microbial action. Efforts to enhance the biodegradability of synthetic polymers, such as blending them with natural polymers or incorporating biodegradable additives, will remain an active area of research.

Table 3 Table summarizing polymer matrices and the microbial species/enzymes involved in their degradation, [1, 17, 26]

Polymer matrix	Microbial species/enzymes
Polyethylene (PE)	Phanerochaete chrysosporium (under nitrogen- and carbon-limited conditions), Pleurotus ostreatus, Aspergillus, Cladosporium, Fusarium, Penicillium, Phanerochaete species, Penicillium simplicissimum, Aspergillus niger, Fusarium sp.
High-density polyethylene (HDPE) and low-density Polyethylene (LDPE)	Penicillium chrysogenum NS10, Penicillium oxalicum NS4
Polypropylene (PP)	Bjerkandera adusta, Lasiodiplodia theobromae, Coriolus versicolor
Polyvinyl chloride (PVC)	Cochliobolus sp., Phanerochaete chrysosporium, Aspergillus niger, Penicillium funiculosum, Trichoderma viride, Paecilomyces variotii, Aureobasidium pullulans, Chaetomium globosum
	Yeast-like fungi: Rhodotorula aurantiaca, Kluyveromyces spp.
Polystyrene (PS)	Phanerochaete chrysosporium, Gloeophyllum trabeum (white-rot and brown-rot fungi), Aspergillus, Penicillium
Polyurethane (PUR)	Aspergillus, Penicillium, Trichoderma, Pestalotiopsis microspora, Comamonas acidovorans, Enzymes: Proteases (more effective than esterases)
Polycarbonate (PC)	Phanerochaete chrysosporium, Geotrichum spp., Enzymes: Aryl alcohol oxidases, Mn^{2+}-oxidizing peroxidases
Other plastics (e.g. BPA, DEHP)	Pleurotus ostreatus, Agaricus bisporus (edible fungi), Enzymes: Laccases

5 Impact of Recycling

5.1 Mechanical Properties

The mechanical performance of polymer composites is a critical factor in determining their recyclability. The impact of reprocessing on the properties of neat polymers is influenced by various factors, such as:

- Recycling technique (e.g. thermomechanical, chemical, aging).
- Processing parameters (e.g. temperature, time, speed, shear rate, humidity, UV exposure).
- Molecular composition of the polymer itself.

The impact of recycling and reprocessing of composite waste into upcycled products is typically evaluated through simulated recycling, which involves repeated thermo-mechanical reprocessing methods such as grinding or shredding, extrusion, melt compounding, compression moulding or injection moulding. In some cases, composites are dissolved in an appropriate solvent, cast into films, and then re-extruded or processed using other techniques. Ageing of NFRCs can be simulated by subjecting the material to prolonged exposure at elevated conditions. The addition of reinforcing agents, such as particles, fibres or other fillers further complicates the relationship between recycling and reprocessing processes and the resulting material performance.

The following sections discuss the effects of recycling and reprocessing on the properties of upcycled polymer composites.

5.1.1 Tensile Properties

The recycling processes affect the tensile properties, including strength, modulus, failure modes, and strain properties of NFRCs. These properties can improve or degrade with successive recycling cycles, depending on factors such as fibre aspect ratio, dispersion, fibre-matrix compatibility, and the molecular weight of matrix polymers. For instance, studies on LDPE composites with waste cotton textile reinforcement showed an initial increase in tensile strength and modulus, followed by a decline due to polymer degradation and chain scission. Fibre orientation also impacted tensile performance, with parallel alignment to extrusion direction yielding higher tensile strength [27, 28].

Similar trends were observed in composites with wood fibre, coriander straw fibre, and other natural fillers, where initial improvements in tensile properties were followed by degradation in subsequent recycling cycles. In wood fibre/HDPE composites, tensile modulus decreases initially due to fibre size reduction and polymer molecular weight decrease, but later improve with polymer crosslinking and increased crystallinity. In case of bagasse fibre composites, recycling may improve the tensile properties due to increased fibre aspect ratio [28].

Recycling can help restore the mechanical performance of aged NFRPCs due to improved fibre dispersion, crystallization, and polymer crosslinking. However, in some composites, such as PLA-based ones with natural fibres, significant reductions in tensile strength occurs due to fibre fracture and matrix degradation as in the case of PLA's hydrolytic degradation at high temperatures. Similar reductions in tensile properties were observed in PP and HDPE-based composites, although some improvements were noted in specific cases due to fibre dispersion, polymer crosslinking, and stress transfer [29].

5.1.2 Flexural Properties

Recycling affects the flexural properties of NFRPCs differently than tensile properties due to the difference in stress distributed during testing. Initial recycling cycles often improves flexural strength, but repeated cycles cause deterioration in flexural strength due to fibre and polymer degradation. For example, composites like PP and PE with coriander straw fibre displayed less decline in flexural properties than in tensile properties but with UV irradiation and thermal cycles in humid conditions, the flexural strength was improved by 30%. Factors such as fibre length reduction, polymer degradation, and adhesion loss at fibre-matrix interface cause performance to decline in NFRPCs over multiple cycles. Bagasse fibre/PP and bamboo fibre/polylactic acid (PLA) composites exhibited initial improvements in flexural modulus due to better fibre dispersion and increased crystallinity, but fibre reduction and polymer degradation eventually led to deterioration. Similarly, poplar fibre/HDPE composites showed a 13% initial increase in flexural strength, followed by an 11% decrease [28, 29].

The extent of changes in flexural properties depends on the specific fibre-polymer system and processing conditions. For example, pine fibre/PVC composites showed a 40% increase in flexural strength after 20 recycling cycles due to improved fibre dispersion and crosslinking in the PVC matrix. On the other hand, wood fibre/HDPE composites displayed mixed results, with some studies showing improved fibre adhesion after limited recycling and others showing a decline after multiple cycles. Not all recycling outcomes are beneficial, as many composites, including those with wood, flax, bagasse, or sisal fibres, show consistent declines in flexural performance due to polymer degradation, fibre fracture, and decreased fibre-matrix adhesion. PLA composites, in particular, experience significant declines due to hydrolytic and thermal degradation of both the matrix and fibres [3].

5.1.3 Impact Strength

Recycling has shown a positive influence on the impact strength of NFRP composites due to the reduction in fibre length during reprocessing which enhances ductility. However, this comes at a cost of decrease in tensile and flexural strength, along with modulus while strain at break increases. The degradation of the fibre-matrix interface during reprocessing also contributes to improved impact properties, as weakened bonds allow fibres to slip, enhancing energy absorption and toughness [28].

The impact strength of recycled NFRPCs is influenced by factors such as fibre type, filler content, and processing conditions. Composites with lower fibre content tend to retain or improve impact strength, while those with higher fibre content may experience a reduction in toughness. Optimizing recycling processes and composite formulations is crucial to balance toughness and mechanical performance.

However, some composite systems experience reduced impact strength after recycling due to high fibre damage and/or high filler content that Favors the shear forces, further damaging the fibres and reduce energy absorption. Optimizing recycling processes and composite formulations is essential to achieve the desired balance between toughness and mechanical performance in recycled NFRPCs.

5.2 Thermal Properties

Recycling NFRPCs through thermomechanical or chemical approaches can significantly affect their thermal properties due to changes in the polymer matrix and fibre structure. Reprocessing typically reduces the polymer's molecular weight and fibre length, resulting in shorter and dispersed fibres, which affects the properties like crystallinity, thermal stability, and glass transition temperatures (Tg).

Recycling often results in lower crystallization temperature, as shorter polymer chains facilitate early crystallization. In composites with high filler content, the cold crystallization exotherm may disappear or lower the required temperature. The Tg and melting temperature (Tm) tend to decrease with recycling, primarily due to increased chain mobility from polymer degradation, although Tm may slightly increase in some cases due to severe fibre degradation [28].

Thermal stability may improve if reprocessing enhances crystallinity or promotes crosslinking between fibres and polymer chains, but severe degradation during recycling can decrease stability. Additionally, recycling increases the coefficient of thermal expansion (CTE) and reduces the heat deflection temperature (HDT) which weakens the composite's ability to resist temperature-induced deformation.

The thermal properties of recycled NFRPCs are complex and influenced by recycling steps, filler content, and processing conditions. Understanding these effects is essential for optimizing the recyclability and performance of NFRPCs in practical applications.

5.3 Degree of Crystallinity

Recycling natural fibre-reinforced polymer composites (NFRPCs) generally enhances their crystallinity due to changes in the polymer matrix and fibre interactions. The increase in crystallinity is largely attributed to the nucleation effect of fibres, which promote crystal growth during cooling. Higher fibre content typically results in greater crystallinity because more nucleation sites are available. However, not all studies report increases in crystallinity. In some cases, excessive recycling or high fibre content can reduce crystallinity. For example, bamboo fibre/PLA and wood fibre/HDPE composites showed decreased crystallinity after recycling due to fibre degradation disrupting the nucleation process. The relationship between fibre content and crystallinity is complex, as lower fibre loadings often enhance crystallinity, while higher fibre content may hinder it due to reduced polymer available for crystallization or fibre interference [28, 30].

Recycling leads to polymer degradation, fibre fracture, and shorter polymer chains, all of which contribute to improved crystallinity. Optimizing fibre content, processing conditions, and recycling cycles is crucial for maintaining or enhancing crystallinity in recycled composites.

5.4 Hygroscopic Behaviour

Recycling can improve the hygroscopic properties of NFRPCs, which are commonly required for outdoor applications. Due to the hydrophilic nature of natural fibres, NFRPCs tend to absorb moisture, leading to issues like fibre swelling, dimensional instability, warping, and reduced mechanical properties. Moisture typically enters through fibre-polymer interfaces, fibre pores, and micro-voids in the matrix [31].

Recycling steps such as extrusion, improves fibre dispersion and its encapsulation within the polymer matrix. This results in reduced water absorption, fibre swelling, and improved interfacial bonding. Moreover, thermal degradation of hydrophilic components like hemicellulose during recycling reduces the material's moisture absorption capacity, leading to increased hydrophobicity. Optimizing factors including fibre type, processing conditions, and recycling cycles are critical to improve the durability of recycled NFRPCs for outdoor applications.

5.5 Viscoelastic and Rheological Properties

Recycling and reprocessing NFRPCs influence their viscoelastic and rheological properties. Initially, reprocessing improves the viscoelastic properties such as an increased elastic modulus due to better fibre dispersion. Reprocessing can also elevate the Tg due to stronger fibre-matrix interactions. However, with prolonged reprocessing, the elastic modulus decreases, and the viscous behaviour increases. This is due to excessive fibre and polymer degradation, micro-void formation, and poor fibre-matrix interaction caused by mechanical stresses in processes such as extrusion or injection moulding.

Recycling and reprocessing also alters the rheological properties, such as viscosity and melt flow index, by reducing the polymer's molecular weight and shortening the fibres. This can enhance short-term processing, but excessive cycles weaken the material's structural integrity and long-term performance.

5.6 Ageing and Durability

The durability of NFRPCs in outdoor applications is influenced by environmental factors like temperature fluctuations, moisture, microbial activity, and UV radiation. Accelerated aging tests, such as freeze–thaw cycling, UV exposure, and long-term natural weathering tests, are used to evaluate the aging performance of NFRPCs over the period of operational life. NFRPCs generally exhibit better weathering resistance than neat plastics due to the UV radiation absorption resistance capability of natural fibres, which protects the polymer matrix.

Long-term UV exposure leads to degradation in NFRPCs which results in increased water absorption, colour changes, and reduced mechanical properties. Composites with lower fibre content tend to degrade more significantly under UV exposure. UV-induced degradation can cause cracking, surface degradation and colour changes, primarily due to the bleaching of lignin in wood fibres. Typically, lignin acts as an antioxidant but prolonged UV exposure can bleach it and yellow the polymer matrix. The molecular weight of the polymer matrix also decreases with UV exposure. The reprocessing the aged composites can somehow restore the lost mechanical properties. The addition of UV stabilizers or absorbers, such as hindered amine light stabilizers, carbon black, or nano-silica, can mitigate the effect of UV and enhance the durability of NFRPCs under UV exposure, improving long-term performance in outdoor environments.

6 Mitigation Approaches

The recycling and reprocessing of NFRPCs often lead to the degradation of both the polymer matrix and the natural fibres reinforcement and eventually the material's mechanical properties. This degradation results from irreversible mechanical, chemical, or irradiative changes that occur during the first processing and service life. Reprocessing can also damage additives used in the original composite, further compromising their performance. However, mitigation strategies can be employed to restore or enhance the overall performance of recycled NFRPCs. Key strategies for improving the properties of recycled NFRPCs are listed in Table 4. The mitigation strategies mainly involve addition of functional additives and blending virgin polymers with the recycled materials.

The addition of functional additives such as compatibilizers, coupling agents, processing aids, and thermal stabilizers can improve the fibre-matrix interface and thermal stability of recycled materials. Compatibilizers and coupling agents improve the interaction between polar natural fibres and non-polar polymer matrices which leads to enhanced fibre dispersion, adhesion, and mechanical properties. Processing aids, such as LiCl salts and natural rubber plasticizers, helps in lowering the polymer's melting point and enables the reprocessing at lower temperatures to help preserve fibre integrity. Thermal stabilizers, including antioxidants and flame retardants that protect the polymer matrix during recycling, prevent oxidative degradation and improve flame resistance. The addition of functional additives, therefore, enhances the overall stability and performance of recycled composites.

Blending virgin polymers into recycled composites helps in the recovery of lost mechanical properties caused by the recycling and re-processing cycles and brings them closer to the properties of virgin materials. Optimizing the mixing ratios of recycled material to virgin polymer can effectively balance cost and performance. As the aspect ratio of fibres in NFRPCs decreases with increasing number of recycling and reprocessing, adding high aspect ratio fillers can mitigate this effect and reinforce

Table 4 Key mitigation approaches, their benefits and challenges to enhance the properties of recycled NFRPCs. [28]

Mitigation approach	Key components	Benefits	Challenges
Addition of virgin polymers	Virgin thermoplastics (e.g., HDPE, PP)	Restores mechanical properties, improves processability	Increased cost, potential incompatibility, reduced environmental impact
Compatibilizers and coupling agents	MAPP, NaOH, silane, maleic anhydride, bio-based agents. Acrylic acid-grafted PP, titanium-derived mixtures, bio-based coupling agents like ionic liquids	Improve fibre-matrix bonding, strength, and toughness	Reduced effectiveness after multiple cycles, potential degradation
Processing aids	LiCl, natural rubber, plasticizers	Lowers processing temperatures, reduces fibre degradation, improve properties	May affect colour, cause reactions, require specific conditions
Thermal stabilizers	Antioxidants, montmorillonite, paraffin-treated sawdust	Prevents degradation and mechanical properties	Can be expensive, additional processing steps, environmental concerns
Chain extenders	Polymeric chain extenders	Restores molecular weight, brittleness and enhances properties	May not be effective in all polymers, compatibility issues, cost
Other methods	Bio-based additives (e.g., starch gum, lignin), fibre surface treatments (e.g., NaOH, alcohol wash)	Improves fibre-matrix adhesion, reduces degradation, enhances properties	Limited research, scalability issues, inconsistent results

the structure. Further research into these strategies is important for the sustainability and circularity of NFRP composites.

7 Applications of Recycled NFRPCs

With the growing emphasis on sustainable practices and the circular economy, recycled and reprocessed NFRPCs present substantial opportunities for diverse applications across a wide range of industries. The recycled NFRPCs offer versatile applications in construction, furniture, transportation, sports, and electrical sectors. Their lightweight, durable, and eco-friendly nature makes them ideal for panels, structural boards, automobile interiors, and acoustic insulation. They also find use in storage

Table 5 Potential applications of reprocessed NFRPCs in various sectors

Industry	Applications
Construction and building	Panels for partitions and suspended ceilings, partition boards, walls, floors, window and door frames, roof tiles, prefabricated buildings, temporary shelters in disaster-prone areas
Storage solutions	Post-boxes, grain storage silos, biogas containers
Furniture	Structural boards, chairs, tables, beds, cupboards, wall units, bathroom units
Electrical and electronics	Electrical appliance casings, pipes
Everyday items	Lampshades, suitcases, brush handles
Transportation	Interior components for automobiles and railway coaches, parts for boats
Toys and recreational goods	Toys, entertainment products
Sports equipment	Tennis rackets, golf club shafts
Acoustic applications	Sound insulation panels

solutions, everyday items, sports equipment, and toys, contributing to sustainability and innovation in material use. The potential applications of reprocessed NFRPCs in various industries are listed in Table 5.

Reprocessed NFRPCs demonstrate sustainability and circularity by repurposing waste materials into high-value applications. Recycling and utilizing natural fibres in composites reduce dependency on virgin materials, lower carbon footprints, and promote resource efficiency. This aligns with circular economy principles, ensuring environmental preservation while enabling innovative, sustainable material use across diverse industries.

8 Challenges and Opportunities

Advancing the recycling of NFRPCs requires a multi-faceted approach, including designing inherently recyclable composites, improving the scalability and efficiency of recycling technologies, and developing innovative manufacturing techniques to integrate recycled materials effectively. Besides that, establishing robust waste management infrastructure while maintaining the quality of recycled materials, and adhering to clear regulations are critical to fostering a structured recycling industry is also crucial. Key to achieving sustainable recycling practices will be enhancing the technological readiness, refining the processes for industrial scalability, and promoting collaboration across the product lifecycle and stakeholders. Moreover, recycling initiatives significantly reduce environmental impact, with life cycle assessments showing up to an 80% reduction in carbon footprint when using recycled instead of virgin fibres. Together, these strategies emphasize the need for systemic advancements to support the circular economy. These challenges with potential research opportunities and future directions are summarized in Table 6.

Table 6 Future directions and research opportunities for recycling natural fibre-reinforced composite materials

No	Key area	Description
1	Enhanced recyclability in composite design	Develop composites with inherent recyclability, enabling easier separation of fibres and matrices during recycling, minimizing material loss, and reducing energy consumption
2	Efficiency and scalability of recycling technologies	Address scalability challenges by advancing technologies for efficient industrial-scale recycling. Focus on reducing energy and cost requirements while ensuring high material recovery rates
3	Innovative manufacturing with recycled materials	Create manufacturing techniques that effectively integrate recycled natural fibres into new products. Overcome challenges like diminished fibre strength and altered matrix compatibility from prior use
4	Robust waste collection and processing Infrastructure	Establish structured waste management systems, including collection networks, efficient sorting technologies, and centralized recycling facilities, to support large-scale recycling of NFRPCs
5	Quality retention in recycled materials	Prioritize methods to ensure that recycled materials retain mechanical and chemical properties comparable to virgin materials, improving their performance and marketability
6	Regulations and norms for a structured recycling industry	Develop clear guidelines and frameworks for material standards, recycling practices, and incentives, fostering a structured and thriving recycling industry
7	Mastery of recycling processes for industrial scalability	Refine existing recycling technologies to enable industrial-scale applications. Focus on automation and workflow optimization to reduce costs and enhance efficiency
8	Technological readiness and maturity	Increase the technological readiness level (TRL) of recycling processes to bridge the gap between experimental methods and large-scale operational systems
9	Collaboration across the product lifecycle	Foster collaboration at every stage of the product lifecycle, from design to disposal, to produce sustainable materials, enhance recycling efficiencies, and reduce waste
10	Environmental impacts of recycling initiatives	Reduce the carbon footprint of composite materials through effective recycling initiatives. For instance, replacing virgin carbon fibres with recycled materials can mitigate up to 80% of a product's carbon footprint, as shown by life cycle assessments

By addressing these challenges, the recycling of natural fibre-reinforced composite materials can transition into a viable, sustainable industry capable of supporting growing demand while reducing environmental burdens.

References

1. Khalid MY, Al Rashid A, Arif ZU, Ahmed W, Arshad H, Zaidi AA (2021) Natural fiber reinforced composites: Sustainable materials for emerging applications. Res Eng 11. https://doi.org/10.1016/j.rineng.2021.100263
2. Zimniewska M, Frydrych I, Mankowski J, Trywianska W (2013) 4—Process and quality control in cultivating natural textile fibres. In: Majumdar A, Das A, Alagirusamy R, Kothari VK (eds) Process control in textile manufacturing. Woodhead Publishing, pp 81–108. https://doi.org/10.1533/9780857095633.2.81
3. Elfaleh I, Abbassi F, Habibi M, Ahmad F, Guedri M, Nasri M, et al (2023) A comprehensive review of natural fibers and their composites: an eco-friendly alternative to conventional materials. Res Eng 19. https://doi.org/10.1016/j.rineng.2023.101271
4. Mahmud S, Hasan KMF, Jahid MA, Mohiuddin K, Zhang R, Zhu J (2021) Comprehensive review on plant fiber-reinforced polymeric biocomposites. J Mater Sci 56:7231–7264. https://doi.org/10.1007/s10853-021-05774-9
5. Maiti S, Islam MR, Uddin MA, Afroj S, Eichhorn SJ, Karim N (2022) Sustainable fiber-reinforced composites: a review. Adv Sustain Syst 6. https://doi.org/10.1002/adsu.202200258
6. Bangash MK (2023) Polymer recycling techniques. In: Shaker K, Hafeez A (eds) Advanced functional polymers: synthesis to applications. Springer Nature Singapore, Singapore, pp 199–216. https://doi.org/10.1007/978-981-99-0787-8_11
7. Park S-J, Seo M-K (2011) Chapter 6—Element and processing. In: Park S-J, Seo M-K, (eds) Interface science and composites, vol 18. Elsevier, pp 431–99. https://doi.org/10.1016/B978-0-12-375049-5.00006-2
8. Islam S, Karim FE, Islam MR, Saeed MA, Alam KA, Khatun MM (2024) Thermoset and thermoplastic polymer composite with date palm fiber and its behavior: a review. SPE Polymers. https://doi.org/10.1002/pls2.10157
9. Kamarudin SH, Mohd Basri MS, Rayung M, Abu F, Ahmad S, Norizan MN, et al (2022) A review on natural fiber reinforced polymer composites (NFRPC) for sustainable industrial applications. Polymers (Basel) 14. https://doi.org/10.3390/polym14173698
10. Kukreja BSJL (n.d.) Composites in the aircraft industry. https://www.appropedia.org/Composites_in_the_Aircraft_Industry
11. Fortune Business Insights (2024) Automotive composites market size, share & industry analysis. https://www.fortunebusinessinsights.com/automotive-composites-market-102711
12. Fleck A (2022) Plastic waste: here's what it could look like by 2060. https://www.weforum.org/stories/2022/07/recycling-efforts-not-enough-to-solve-plastic-waste-problem/
13. Alves B (2024) Plastic waste worldwide–statistics & facts. https://www.statista.com/topics/5401/global-plastic-waste/#:~:text=between%201970%20to%202019%2C%20an,On%20marine%20life%20and%20ecosystems
14. Qureshi J (2022) A review of recycling methods for fibre reinforced polymer composites. Sustainability (Switzerland) 14. https://doi.org/10.3390/su142416855
15. Gonçalves RM, Martinho A, Oliveira JP (2022) Recycling of reinforced glass fibers waste: current status. Materials 15. https://doi.org/10.3390/ma15041596
16. EU regulations & legislation (2024) Landfill Directive–Directive (EU) 2018/850. https://circular-cities-and-regions.ec.europa.eu/support-materials/eu-regulations-legislation/landfill-directive-directive-eu-2018850

17. Damayanti D, Saputri DR, Marpaung DSS, Yusupandi F, Sanjaya A, Simbolon YM, et al (2022) Current prospects for plastic waste treatment. Polymers (Basel) 14. https://doi.org/10.3390/polym14153133
18. Pakdel E, Kashi S, Varley R, Wang X (2021) Recent progress in recycling carbon fibre reinforced composites and dry carbon fibre wastes. Resour Conserv Recycl 166. https://doi.org/10.1016/j.resconrec.2020.105340
19. Ren Y, Xu L, Shang X, Shen Z, Fu R, Li W et al (2022) Evaluation of mechanical properties and pyrolysis products of carbon fibers recycled by microwave pyrolysis. ACS Omega 7:13529–13537. https://doi.org/10.1021/acsomega.1c06652
20. Ziemińska-Stolarska A, Sobulska M, Pietrzak M, Zbiciński I (2024) A review of end-of-life scenarios for fibre-reinforced polymer materials. Energies (Basel) 17. https://doi.org/10.3390/en17153713
21. Thiounn T, Smith RC (2020) Advances and approaches for chemical recycling of plastic waste. J Polym Sci 58:1347–1364. https://doi.org/10.1002/pol.20190261
22. Seabra CP, Sousa AC, Bragancą IMF, Silva CMA, Robalo MP, Loja MAR, et al (2020) On the performance and recyclability of a green composite based on AESO resin. J Manuf Mater Proc 4. https://doi.org/10.3390/jmmp4030065
23. Benyathiar P, Kumar P, Carpenter G, Brace J, Mishra DK (2022) Polyethylene terephthalate (PET) bottle-to-bottle recycling for the beverage industry: a review. Polymers 14. https://doi.org/10.3390/polym14122366
24. Recent Development of Biodegradation Techniques of Polymer (n.d.)
25. Silva RRA, Marques CS, Arruda TR, Teixeira SC, de Oliveira TV (2023) Biodegradation of polymers: stages, measurement, standards and prospects. Macromol 3:371–399. https://doi.org/10.3390/macromol3020023
26. Ciuffi B, Fratini E, Rosi L (2024) Plastic pretreatment: the key for efficient enzymatic and biodegradation processes. Polym Degrad Stab 222. https://doi.org/10.1016/j.polymdegradstab.2024.110698
27. Sonmez HE, Berkalp ÖB, Bakkal M, Bodur MS (2014) Effect of recycled LDPE matrix on the properties of waste cotton fiber reinforced (WCF-R) composites. Usak University J Mater Sci 3:135. https://doi.org/10.12748/uujms.201416507
28. Zhao X, Copenhaver K, Wang L, Korey M, Gardner DJ, Li K, et al (2022) Recycling of natural fiber composites: challenges and opportunities. Resour Conserv Recycl 177. https://doi.org/10.1016/j.resconrec.2021.105962
29. Mukherjee T, Kao N (2011) PLA based biopolymer reinforced with natural fibre: a review. J Polym Environ 19:714–725. https://doi.org/10.1007/s10924-011-0320-6
30. Long H, Wu Z, Dong Q, Shen Y, Zhou W, Luo Y et al (2019) Mechanical and thermal properties of bamboo fiber reinforced polypropylene/polylactic acid composites for 3D printing. Polym Eng Sci 59:E247–E260. https://doi.org/10.1002/pen.25043
31. Tajvidi M, Takemura A (2010) Recycled natural fiber polypropylene composites: water absorption/desorption kinetics and dimensional stability. J Polym Environ 18:500–509. https://doi.org/10.1007/s10924-010-0215-y

Recycled Bio-waste Derived Fiber-Reinforced Composites (BFRC) and Their Applications

Rajesh Kumar Mishra, Shabnam Nazari, and Tatiana Alexiou Ivanova

Abstract This chapter summarizes the emerging use of bio-wastes such as rice husks, corn stalks, and sugarcane bagasse etc. coming from plants' waste for reinforcement is one of the most accepted ways to bring sustainability in fiber reinforced polymer composites. The biowastes are discussed from the point of view of their properties and compositions that allow them to be used as reinforcement in polymer composites. This chapter discusses the recovery, recycling, and utilization of the wastes arising from this category of composite materials. The chapter is focused on the utilization of residual agricultural biomass/natural fiber waste as a promising sustainable construction material in rural areas. The natural/agro based fibers suitable for use in sustainable construction materials are summarized. The effect of fibrous reinforcement on the mechanical, thermal, acoustic and fatigue performance of sustainable construction materials are described. This chapter further addresses the technologies, techniques, challenges, and opportunities linked to the recycling of fiber reinforced polymer composites.

Keywords Biomass · Bio-waste · Fiber reinforced polymer composites · Recycling · Reuse · Applications

1 Introduction

Today, energy crisis is an important and problematic issue in the world. Population expansion, industrialization, and improved living standards have contributed to a rise in energy use. As a result, energy is considered a valuable commodity, and all governments are looking for a means to gain access to more energy sources for a variety of needs while spending as little as possible. One of the essential sources of energy that has a special place today is fossil energy. Researchers are looking for solutions to reduce energy usage in numerous fields as a result of the high cost of pollution caused by this energy and humanity's failure to locate alternative fossil fuels. Efficient use of

R. K. Mishra (✉) · S. Nazari · T. A. Ivanova
Czech University of Life Sciences Prague, Prague, Czechia, Czech Republic
e-mail: mishrar@tf.czu.cz

H. Jamshaid and U. Zubair (eds.), *Sustainable Recycling of Fiber Reinforced Polymer Composites*, SDGs and Textiles, https://doi.org/10.1007/978-981-96-9536-2_3

thermal energy by using insulation materials can reduce fossil fuel use. Thermal insulation is beneficial for various reasons; the principal one is to save energy. Thermal insulation can reduce heat loss and hence the amount of energy required to maintain the process fluid temperature, resulting in cost savings [1, 2]. However, for several reasons, the high cost of heat insulation versus energy consumption reduction is very important. It is hoped that by conducting this research an effective step can be taken in the field of application of thermal insulation. Because of possible renewability, relatively lower density, improved specific mechanical performance, biodegradability, accessibility, and affordability, fibers are attractive in composite materials. In contrast to man-made fibers, biofiber reinforced composite materials offer superior mechanical performance at a lower cost. Currently, a number of biobased green composites are developed with improved mechanical properties by using different natural origin fibers and biomass. The purpose of such research is to develop a product with a small volume that is a biomass-based three-dimensional (3D) hybrid composite [3–5]. Reinforcement by biomass or biofibers can have excellent properties, such as thermal, moisture, and acoustic insulation; it also provides a large part of the desired mechanical properties (including bending) and can be made environmentally friendly.

Using efficient heat insulation is considered as the most efficient energy saving option in houses and buildings. The resistance with a layer of insulation material depends critically on the thickness and thus decreases thermal conductivity. Under conditions of dynamic heat flow, (as in many practical fields), heat insulation materials also effect several other related characteristics such as the heat decrement factor, the delay of time, and the peak load of heat transmission. Many different types of heat insulation materials are currently available, with differing thermal, mechanical properties and the cost. Based on the overall structure of walls, pillars, beams and roofs, diverse insulating practices are available. The thermal performance in case of polyester resin based composites reinforced with bamboo viscose fibers was investigated. It was found that the thermal conductivity in case of fiber-reinforced composites varies with the change in fiber volume fraction, the orientation angle of the constituent fibers, and the outside temperature [6].

The purpose of conserving energy is not just the reduction of expenses but also the conservation of precious natural resources for use in the future. Fuel consumption is 80% based on non-renewable energy sources. Therefore, it is utmost important for many industries such as housing and construction, transportation and aerospace to practice methods to conserve energy [7–9]. Composites and polymer based materials contribute to the achievement of this purpose. The main characteristics of the polymer composites are relatively lower density, higher specific mechanical strength, higher specific stiffness/rigidity, and superior surface hardness properties [10, 11]. Natural cellulosic fibers are extensively used as reinforcement in composites from thermoplastic and thermoset polymers. Recently, there is extensive research on new types of materials having superior performance at lower costs. Increase in environmental awareness also led to further focus on eco-friendly materials which are renewable, possibly recyclable, and a sustainable option. Polymer composites, prepared by using natural or man-made reinforcements and a wide variety of matrices, are included in this approach. Lignocellulosic fibers like jute, agave, sisal, hemp, pineapple, curauá, etc. can be used to improve employment opportunities in rural areas [12, 13]. The extraction of such fibers is a very important step which determines their performance. They have proven to be an effective reinforcement filler in case of several polymers, and composites using these fibers have become increasingly popular on industrial scale especially in the modern world [14, 15]. In addition to their softness, diverse color, high cost value, versatility, natural luster, uniformity of size, and desired length, jute is a preferred fiber in several composite applications [16].

The heat conductivity of 3D structures based on natural fibers has been studied and contrasted with those of their synthetic counterparts. Building structures should be more resistant to heat flow and the heat transfer capacity of the composite material is crucial in determining how responsive it is to thermal energy. Building materials should also have properties of thermal stability, thermal expansion, and thermal flammability. Application success depends on an understanding of the relative expansion/contraction properties of the materials. From the literature review, until now, a number of research workers have studied the effect of hybridization with different natural as well as man-made fibers for applications in composites. The investigations on 3D jute fabric/waste fiber in a hybrid composite is rare. Composites are materials that consist of two or more types of distinct materials with different physical and chemical properties. These materials/phases are separated by an interfacial layer. Composite materials can be classified into particulate filler based, fibrous structure based, and laminate composites mainly based on the types of reinforcement being used. Particulate filled composites are reinforced by particles of various nature, fibrous composites are reinforced by fibers and their geometrical structure, and laminated composites are single or multilayered. Fibrous structure based composites can be reinforced with natural or man-made fibers. The natural fiber reinforced composites are also known as biofiber composites or sometimes simply as biocomposites. There exist several types of natural resins which can also be classified into non-biodegradable or biodegradable varieties. Combination of natural fibers and biodegradable polymer resins can produce composites popularly known as purely biobased or green composites. These can be further classified into hybrid and purely green composites [17, 18]. Hybrid composites are made from the mixture of two or sometimes more varieties of fibers. Due to their superior corrosion resistance, high strength-to-weight ratio, simple production process, and long fatigue life, polymer composite are attracting a lot of research attention lately. They are mostly useful in the automotive, civil engineering, and aviation industries. Infrastructure and construction industry accounts for 42% of all composite material usage, making it the largest market for composites among all other industries [19].

There are numerous examples of the use of materials from biological sources for making components used in automotives. Beyond conventional materials like wood, cotton fibers, textile based materials, and leather based seats, there can be several applications: as reinforcing fibrous structures or fillers, or even polymer nanocomposites. Plant based fibers (like hemp, sisal, flax, kenaf and jute) can be used as reinforcing fibers. Biopolymers can be derived from various sources, including soy, castor oil, corn, or sugarcane. These feedstocks are typically fermented and undergo a sequence of conversion processes for producing polymer resin composites. Biopolymers can also be extruded, molded, expanded, and thermoformed like conventional polymers. Several articles as interior door panels, backset shells, armrests in chairs, and floor tiles/panels can be made from sisal or hemp fibers. Fibers from coconut and biofoams are used to make seats, headrests on chairs, and handrests in automotive seats. Several natural origin fibers have potential to provide excellent sound insulation and therefore, they are traditionally used as covers of interior walls in buildings [20].

Bio-based composites and derived articles are generally used in several ways e.g., to replace plastic/polymer components made from petroleum or to replace man-made fibers (like carbon, Kevlar, glass fibers). Use of nature-based products which are biodegradable can help in reducing carbon emissions by elimination of hazardous chemicals and their products. Polymer composites based on biomaterials are lighter and less dense as compared to man-made counterparts. Bio-based composite materials can absorb impact energy, while mineral fiber-based composites lead to fracture. Therefore, the aforementioned materials are safer for occupants in a vehicle in case a road accident occurs. It has been reported that the natural fiber-based composites require much lower energy (about 11.4 MJ/kg in average) during the whole process of manufacturing in comparison with glass fibers which is about (48.3 MJ/kg in average) [21].

2 Composites from Agricultural Biomass

2.1 Agro Waste and Recycled Wood Waste in Composites

It is a novel approach to use alternative raw materials from agricultural biomass as well as recycled waste of wood and related by-products for producing particleboards. This helps to solve the increased demand worldwide for utilization of wooden materials. This also helps in maintaining the principles of circular economy. Wooden fragments are also very costly element like the polymer resin used in the production of composites. Both elements account for more than half of the total production cost. So, by replacing wooden fragments with lignocellulosic agricultural waste-based biomass can save a significant amount of cost. Such biomass from agriculture is available abundantly during the post-harvesting and post-production operations and they can be used as ideal alternative for preparing particleboards. Extensive research has been conducted for evaluation of the current methods of producing composites by use of a large variety of ecologically friendly agricultural waste and biomass, recycled waste of wood, and related by-products. The biomass from agricultural waste used for producing composites can be classified into several different types based on the plant type and their parts which can be extracted from, i.e. stems, straw, stalk, skin, bark, bagasse, seed/fruit, leaves, grass, and palm fruits etc. as shown in Fig. 1 [22]. The characteristics of such composite materials depend mainly on the

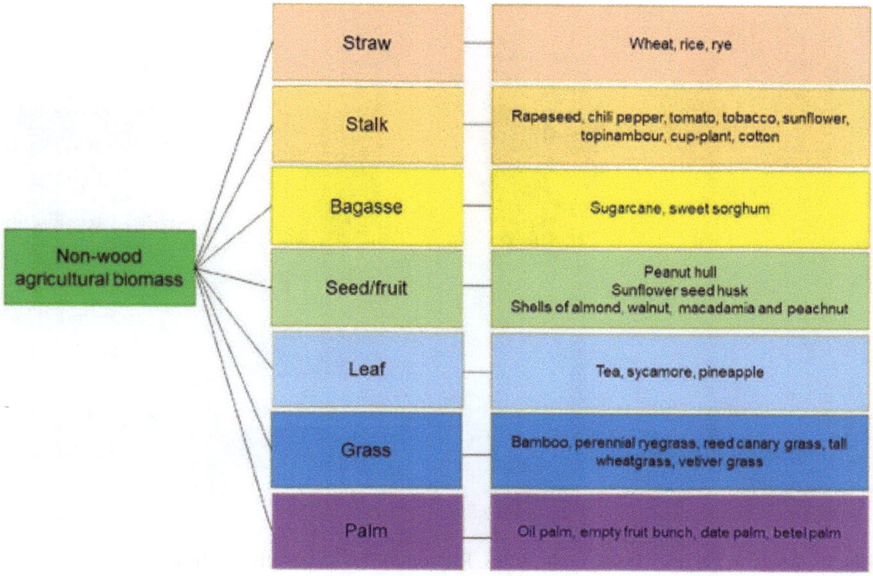

Fig. 1 Classification of non-wood biomass. Reproduced with permission from [22]

type of raw material used and their mechanical properties. There are several challenges and potential for future in using biomass from agricultural waste as well as recycled waste from wood [22].

2.2 *Preparation of Biomass Composites from Waste Wood Fibers*

The ecosystem in the forests play an important role in controlling global warming, as wood is considered as a vital nature-based material for capturing carbon. The increase in demand for wood worldwide is opposed to the concept of sustainable capture of carbon, emphasizing the importance of waste wood recycling and its utilization it a variety of ways. Though the wood-based fibers have a potential to improve strength in composites, the process of their preparation is very complex. The natural sugar in wood restricts use of fibers derived from wood in several applications. A possible recycling of waste wood is shown in Fig. 2 [23]. Many research activities dealt with the combination of the Random Forest and NSGA II theories for comprehensive optimization using multi-objective methods. The approach focuses on enhancement of strength, developing resilience, and capabilities of capturing carbon by use of composites. Detailed microstructural analyses, mapping of elements, reveal that wood-based fibers can contribute to increase toughness and for higher cracking resistance, and thus it affirms better mechanical

Fig. 2 Waste wood fiber-based composites. Reproduced with permission from [23]

properties of polymer composites. The Pareto solution was derived from the several algorithms developed and the multi-objective problem was solved. Wood fiber based cementitious composites can be created which show several advantages in terms of sustainable product development, and transition of theoretical methods into environment-friendly practical solutions [23].

2.3 *Recycling Non-Food Waste and Biomass into Value Added Composite Products*

Lignocellulose-based biomass (LCB) is understood to be one of the most abundant resources available on the earth. The complex chemical structure of LCB results in its high resistance which is a hindrance to its exploitation in composites effectively. Research has identified the tremendous potential of bio-enzymes in converting LCB agricultural non-food into products of high value through sustainable and eco-friendly procedures. The complex structure and functionalities of several types of enzymes, like lytic polysaccharide monooxygenases (LPMOs), lignin-degrading biocatalysts, hemicelluloses, cellulases, etc. are examined thoroughly and deeply investigated. The enzymes exhibit synergistic actions, and it is very important for successfully breaking down of the LCB structure which can facilitate the release of sugars that are fermentable. This further leads to the production of important by-products like biogas, xylitol, and bioplastics. The extraction of strong and stable cellulose from LCB which is catalyzed by enzymes is also highly profitable because of its possible application in the paper and pulp industry [24, 25].

2.4 Sustainable Composites from Fertilizer-Grown Marine Filaments

The demand for petroleum-based polymers has increased worldwide, and it results in huge amount of environmental pollution due to their consumption and post-consumer waste dumped as landfill. Thermomechanical method to recycle the polymer waste has several advantageous, but it cannot solve all the challenges related to the demand of plastic pollution all over the world. A rather more sustainable and ecofriendly approach for recycling waste plastic can be the production of thermoplastic composite blends by combining waste plastic and alternative biobased materials such as marine biomass derived from algae. Marine cyanobacterium can be cultivated using waste nitrogen-based fertilizer as a source of nitrogen, which results in production of phycocyanin and biomass having a density of 6.5% and 0.7 g/L, respectively. The range of tensile strengths in thermoplastic blended polymers containing biomass, recycled plasticizer from glycerol, and plastic waste can be between 0.29 and23.2 MPa. The tensile strength and elastic modulus of thermoplastic composites can decrease with increasing concentration of the biomass. Further, the thermal analysis can be done to ensure that thermoplastic composites containing biomass can have lower thermal onset and lower degradation temperatures as compared to polyethylene waste. It should also be understood that 35–50% of biomass can be a very sustainable alternative using feedstock from biobased resources in order to produce blended thermoplastic polymers, thus making the recycling process of plastic waste a more sustainable and ecofriendly approach [26–28].

2.5 Wood-Plastic Composite with Recycled Materials

Wood-based polymer composites (WPCs) can be prepared by using recycled waste of expanded polystyrene (EPS) and wood-based sawdust as a biomass material. The recycled vehicle exhaust soot (RVS) was also used as a potential carbon rich additive in preparing the composites. The effect of carbon dust on the properties of the recycled composites was extensively studied. In such process, the carbon soot content was varied in the range of 0–2.5 by wt%. Figure 3 shows examples of wood-based polymer composites using recycled materials.

The recycled composites were developed by using a compression molding method. The influence of recycled carbon soot on the physical, mechanical and thermal properties e.g., tensile modulus, flexural rigidity, compressional strength, impact performance, surface and bulk hardness, power of holding a screw holding/ force required for withdrawal of a screw, water absorption and thermal stability of the prepared WPCs were studied. It was found that the mechanical performance of the composites significantly improved due to the presence of carbon soot and its % loading. The soot also offered beneficial results in improving mechanical properties when it was added to a loading of 1.0 wt% in the recycled composites. However,

Fig. 3 Wood-plastic composite with recycled materials. Reproduced with permission from [29]

the content of carbon soot above 1.0 wt% may result in negative results in the properties of the composites. The possible reason for deterioration in the performance of the composite by adding excess of the soot beyond 1.0 wt% can be because of the agglomeration of carbon soot particles in the network of the polymer composite. The water absorption capacity of the composites continuously decreased with the increasing load of carbon soot from 0 to 2.5 wt% as the carbon soot is hydrophobic in nature. The composites developed are thermally stable till a temperature of about 240 °C [29]. It can therefore be inferred that the recycled carbon soot can be used as a sustainable substitute to the more hazardous commercial carbon material in making polymer composites for future in order to improve its properties while preparing cost-effective and sustainable alternatives for industrial composites.

2.6 *Composites with Magnetic Properties Prepared by Using Humic Acid Derived from Rice Straw-Based Biomass and Waste*

Bio-based humic acid (BHA) was derived from rice-straw. It is rich in several functional groups, as it is obtained from waste from rice straw in a single-step hydrothermal process. The BHA is further co-precipitated with nanoparticles (MNP) having magnetic properties in order to produce novel composites for the adsorption of Cd2 + ion, Pb2 + ion and Ni2 + ion from water resources e.g., rivers, ponds and ocean. Adsorption data from the experiments showed that the composites having magnetic properties can adsorb ions of heavy metal efficiently, with maximum adsorption (Qm) of 290.81, 374.99 and 296.85 $mg.g^{-1}$ for Cd2 + ion, Pb2 + ion and Ni2 + ions respectively. Adding BHA to MNPs can improve their dispersion in solvents and thus the resistance to oxidation inside water. This provides enhanced

Fig. 4 Magnetic composites from rice straw-based biomass. Reproduced with permission from [30]

adsorption points and ultimately increases the specific surface area. Adsorption mechanisms of Cd2 + ion, Pb2 + ion and Ni2 + ion by the magnetically enabled composite are governed by the principle of chemosorption. It is a spontaneous and typically endothermic process. The composites with magnetic properties prepared by this method shown in Fig. 4 are easily separated and can be recycled from solutions. They have excellent adsorption performance and can be regenerated as well. This is a promising and novel green adsorbent composite material for removal of ionic pollutants of heavy metals from water bodies [30].

2.7 *Lignocellulosic Biomass Feedstock for Biodegradable and Recyclable Composites*

Figure 5 shows composite production technologies from Lignocellulosic biomass [31].

A complete life cycle of biomass composites is depicted in Fig. 6.

Polymer-based sugars entrapped in lignocellulose-based biomass (LCB) are extracted by using an adequate pretreatment process for transforming it into fuels, special chemicals, and composite materials. Several renewable and monomeric chemicals are produced from LCBs and they can subsequently be converted into composites which are biodegradable and in many cases recyclable. Such composites can be used as a replacement of traditionally used single-use polymers/plastics apart from making biofuels and biochemical derivatives. Such unconventional methods can have several unique advantages of fractioning a defined portion of LCB to be used in developing high value products in industrial applications. Figure 7 shows all degradable components in LCB.

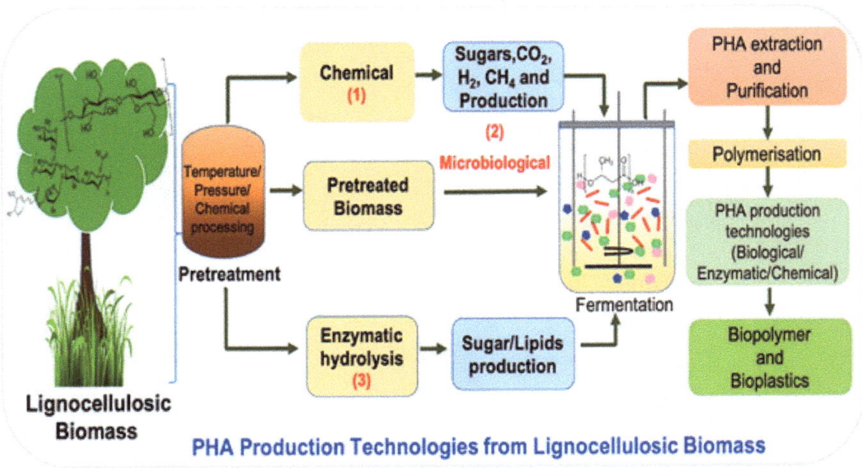

Fig. 5 Production technologies from Lignocellulosic biomass. Reproduced with permission from [31]

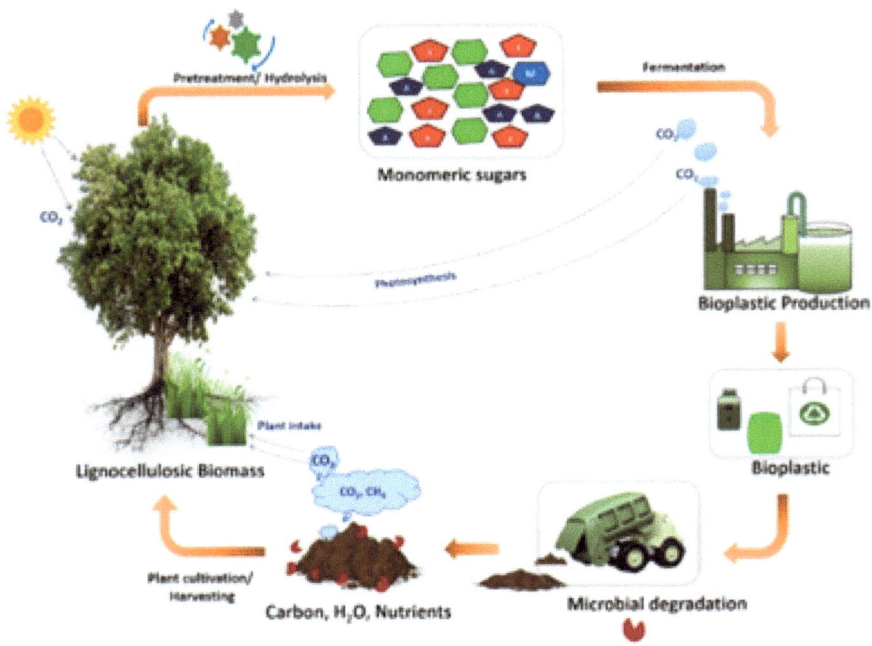

Fig. 6 Complete life cycle of biomass composites. Reproduced with permission from [31]

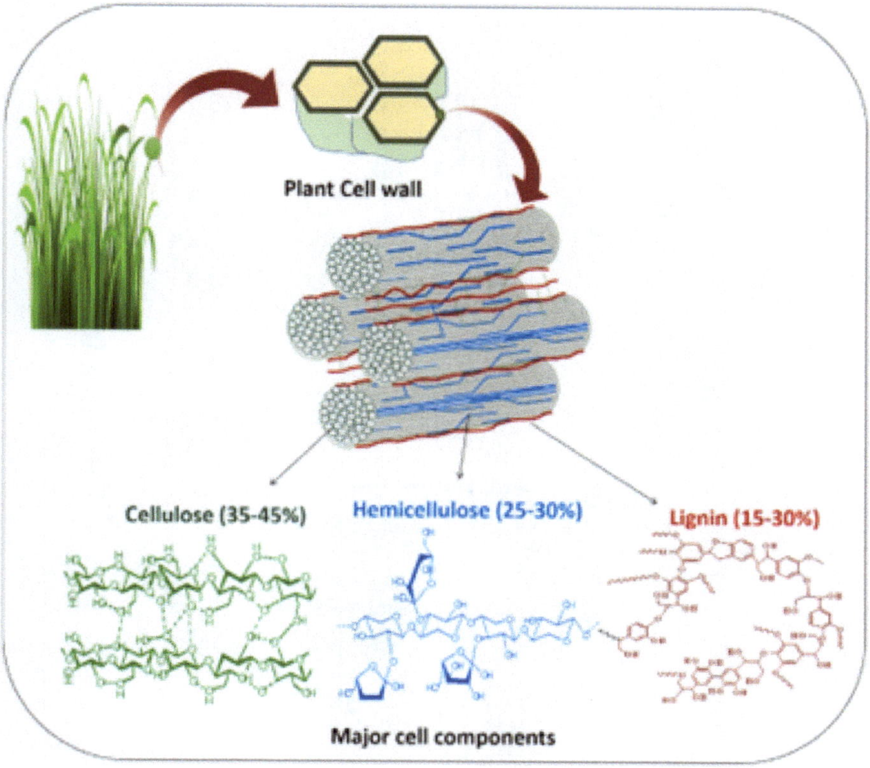

Fig. 7 All degradable components in LCB. Reproduced with permission from [31]

Several polymers such as PHA, PLA, and PGA which are biodegradable, should be selected precisely in the present situation in order to prepare naturally bio-degradable composite materials. Such materials can be used in nano-biomedical, applications in tissue engineering, and several light weight household applications. The use of LCB, which is a resource for recyclable composites, can help in promoting sustainability and concept of circular economy [32, 33]. The variability in the structure, higher level of heterogeneity, more complexity of the LCBs are some of the important aspects which require advancements in research and development activities in both technological as well as commercial direction. This problem can be solved by developing adequate strategies to improve the efficiency of converting the substrate into the final product and thereby improving its commercial viability. Various methodologies for biomass processing are shown in Fig. 8.

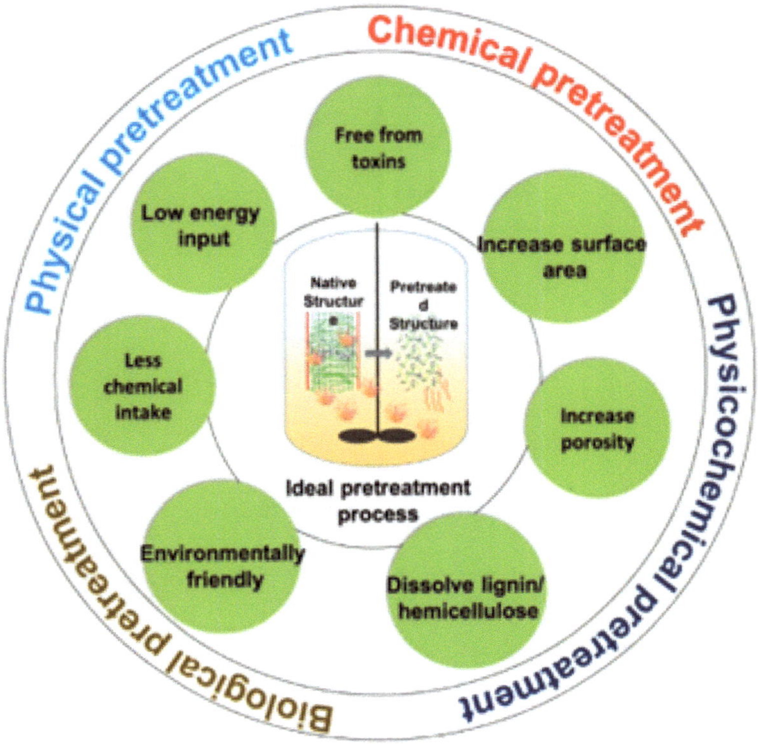

Fig. 8 Various methodologies for biomass processing. Reproduced with permission from [31]

3 Composites from Consumer Waste of Biological Origin

3.1 Upcycling of Polymer Wastes and Biomass into Robust Composites for Energy-Harvesting

The pollution caused by polymeric waste is increasing globally and causing widespread concern. From the viewpoint of decarbonization during the entire life-cycle of the polymers, the development of robust and recyclable composite materials from sustainable resources is a very highly desirable solution. A new supramolecular strategy was developed to fabricate sustainable composite materials by integrating recycled glass fibers and fabrics (RGFF, especially recovered from wastes of wind turbine blades) and bio-based soybean oil (ESO) vitrimer obtained through the epoxidization process [34–36]. The key principle of the process design is the development of a supramolecular interface with the heavily dense and aggregated hydrogen bonds joining the resin material onto the RGFF and ESO vitrimer. This helps in the in situ assembly process in order to form a regular and dense inter-woven structure like cartilage. The materials developed show excellent stability during service

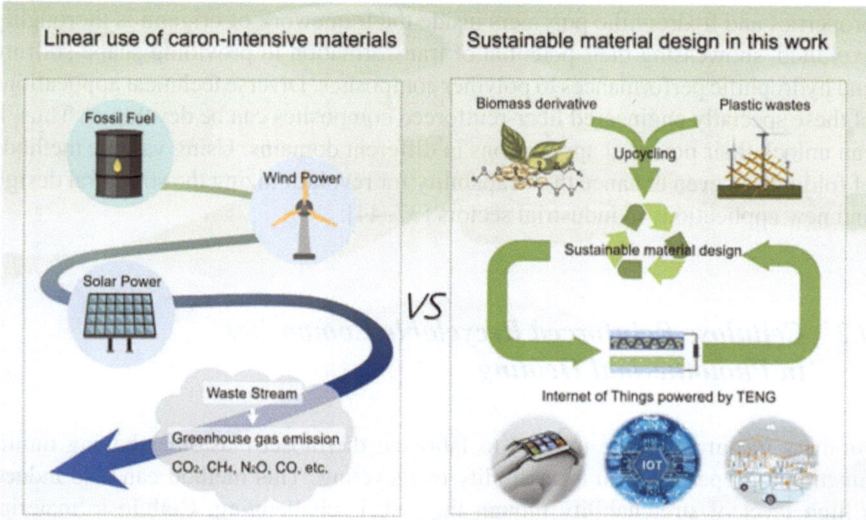

Fig. 9 Biomass based energy-harvesting composites. Reproduced with permission from [36]

life (mechanical strength of 152.9 MPa and bulk toughness of about 33.9 $MJ.m^{-3}$). It also exhibited recyclability after end-of-life. The dielectric modulation was also adequate (promote 11.6 times in surface potential). This allows the developed materials for use as an efficient and durable material for harvesting triboelectric energy, as shown in Fig. 9. The developed strategy for upcycling polymeric waste materials or biomass showed about 44%–49% reduction in the carbon footprint. This further opens up new avenues for developing sustainable composite materials which are potential alternatives to petroleum-based polymers [36].

4 Composites from Industrial Waste of Biological Origin

4.1 Origami-Inspired Fiber-Reinforced Composites

Fiber-reinforced composites which are inspired by the Japanese Origami can exhibit inherent performance properties which highlight their extraordinary response with respect to folding based on external stimuli. Fabrication techniques inspired by the Origami have been meticulously used in fiber-reinforced composites, emphasizing on kirigami and Miura-ori patterns of folding, and several other patterns like waterbomb, Kresling and Yoshimura etc. It highlights very special methodologies of manufacturing, and exploiting advanced techniques like electrospinning, additive manufacturing, and most 4D printing techniques to create meticulously engineered reinforcing structures. The incorporation of polymers having shape memory properties and hydrophilic polymers inside the framework of origami is thoroughly exploited, showcasing their potential of transformation in providing shape-shifting and hydrophilic performances to polymer composites. Diverse technical applications of these specially engineered fiber-reinforced composites can be developed. Thus, it can unlock their potential applications in different domains. Using various methods of folding can even enhance their capability for revolutionizing the structural design and new applications in industrial sectors [37–44].

4.2 Cellulose Reinforced Recyclable Composites in Photothermal Heating

Biomass resource can be utilized to fabricate thermoset elastomer having multi-functional properties with a possibility of recycling. This method can also induce a high level of stretchability though the task is challenging. Cellulosic material reinforced thermoset composite using elastomeric materials having excellent potential for recycling and stretchability can be developed to address such a challenge as shown in Fig. 10. In order to prepare the thermoset polymer elastomers, the cellulose-based macro-molecular crosslinker having maleimide functional groups are utilized to further react with the poly(styrene-block-butadiene-block-styrene) (SBS) through a thermally reversible Diels–Alder (DA) type reaction which can construct the dynamic and highly crosslinked thermoset using elastomers. The creation of a dynamic and highly crosslinked network is further demonstrated by stress relaxation performance, high swelling, and excellent DMA properties. The formation of the cross-linking network can significantly improve the mechanical properties and enable the thermoset elastomer for a possibility of thermal recyclability. The polymer exhibits solid-state plasticity with a high efficiency of recycling around 88–90%. The recyclable elastomeric thermoset polymer composite using carbon nanotubes (CNT) can exhibit excellent photothermal conversion and thermal heating performance. The surface temperature of the composites can reach as high as 126.3 °C with a voltage of 7.5 V by thermal processes. Such elastomeric thermoset composites can be developed for industrial applications like photoelectric generators and conductive coating on other materials [45].

4.3 Adsorbent Composite Materials Using Superparamagnetic Nanoparticles by Using Activated Carbon Derived from Biomass Sludge

Activated carbon magnetic nanoparticles derived from biomass sludge can be used in composites as depicted in Fig. 11 [46]. The overall surface area, volume porosity and pore size can be evaluated by using BET analysis. The adsorbent composite can show

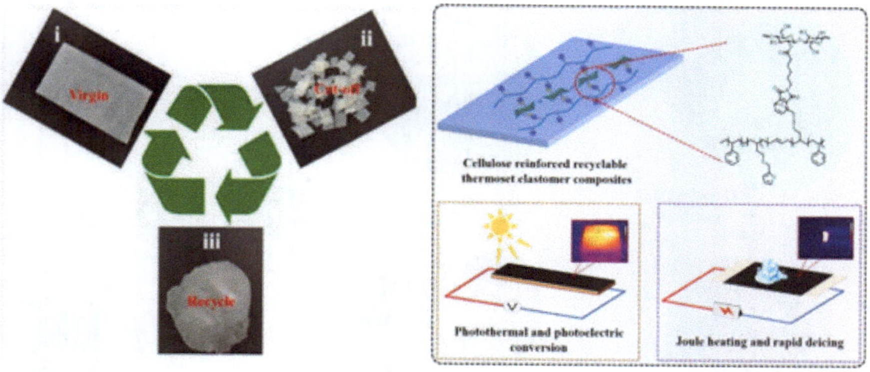

Fig. 10 Cellulose reinforced recyclable thermoset elastomer composite. Reproduced with permission from [45]

a specific surface area as high as119.56 m^2/g and a magnetization value as high as 2.35 emu/g. The size of iron oxide nanoparticles can vary in the range of 15–30 nm. The adsorbent composite material can be utilized for the removal of Chromium (VI) from various aqueous solutions. Isotherm models e.g., Langmuir, Freundlich and Temkin models can be employed to evaluate the phenomena of adsorption. Adsorption capacity of the monolayer adsorbent for Cr (VI) can be as high as 80.21 mg/g at a pH of 2.0. Thermodynamic analysis of the adsorption process of the prepared adsorbent composite shows excellent stability and possibility of recycling for multiple number of uses [46].

4.4 *Chemically Crosslinked Biomass-Derived Green Polymer Composites for Soft Electronics*

Development of green polymer composites is important for overcoming the limitations associated with recycling and reprocessing of several electronic systems. Polymers extracted from α-lipoic acid (or lipoate), which is a bio-derived molecule, is considered to be a promising candidate for developing sustainable alternatives. It offers the advantage of chemical degradation through the process of disulfide reduction or by depolymerization when required. The kinetics of depolymerizing poly(ethyl lipoate) (PEtLp) using different types of acidic and alkaline conditions were observed to be significantly efficient. The solution in chloroform or toluene (20 mg/mL) exhibited full depolymerization (100% conversion) within two days by using trifluoroacetic acid (TFA) at a temperature of 25 °C. The recyclable polymer composite as shown in Fig. 12 can be produced which has the potential for 100% decomposition into a liquid solution. This requires a monomer, a crosslinker, conducting fillers like carbon nanotubes (CNT) which enable conductivity and repeated recycling by chemical methods. The resultant conductive composite exhibits

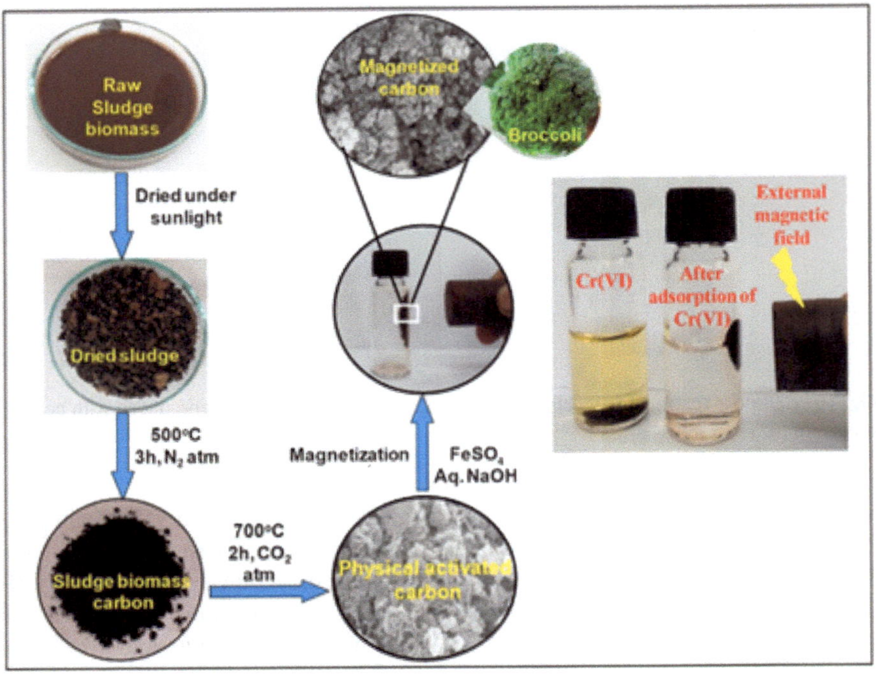

Fig. 11 Superparamagnetic nanoparticles/sludge biomass composite. Reproduced with permission from [46]

extremely low change in electrical resistance (ΔR/R0 ~ 0.15) while showing a stretchability of 100% under uniaxial loading. It also maintains a consistency in the mechanical as well as electrical properties during repeated recycling and reuse under different levels of tensile strains. Such recyclable polymer composites show significant potential as conducting green electrodes used in soft electronic applications [47].

4.5 Humic Acid Based Hydrothermal Carbon Composite Derived from Biomass for High-Efficiency Adsorption

Humic acid (AHA) based hydrothermal carbon (HC) composites can be prepared using biomass materials such as corn straw. Hydrothermal carbon (HC) is used as the matrix to develop high-efficiency adsorption for norfloxacin (NOR). The composites were developed by using a coprecipitation method with the mass ratio of mAHA:mHC = 1:4. This is an environmentally friendly and green method of developing sustainable composites. The oxygen content of such composites can be increased by using an ultrasonic method and a combined acidic-alkaline hydrothermal method. The capacity for saturated adsorption of NOR can be as high

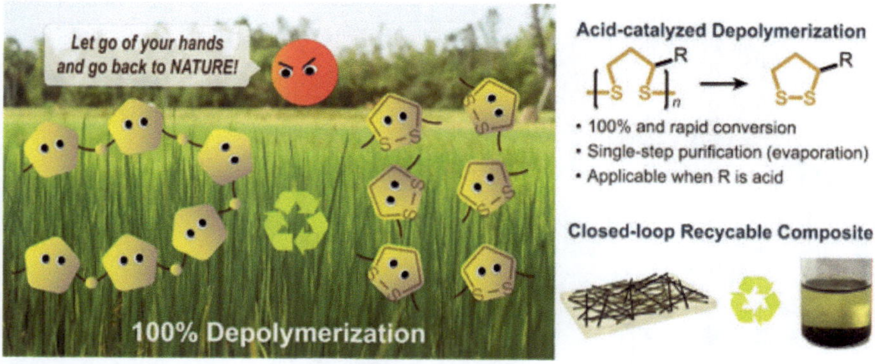

Fig. 12 Biomass-derived closed-loop recyclable composites. Reproduced with permission from [47]

as 565.48 mg/g. This is much higher than several other adsorption materials reported. The high-efficiency of adsorption is because HC is combined with AHA which is rich in functional groups (−OH, −C=O, etc.) containing oxygen. The composites show excellent selective adsorption for NOR, with lower risk from possible acidity of the environmental water. It can show high detection rates for several inorganic ions as well as antibiotics. After several reuse, the residual adsorption capacity can be still higher than 90%. The adsorption mechanism is based on π-π stacking, hydrogen bonding and sometimes electrostatic bonding. The removal rate of NOR from simulated water having a concentration of 10.0 mg/L varies between 91 and 95%, while the concentration is about 16 times of the maximum concentration of 630 μg/L in highly polluted water. It proves that AHA/HC composites can remove all NOR in polluted environmental water with an adsorption capacity far higher than the level of NOR in the environment. Such materials are of high practical significance for recycling of agricultural wastes and for efficient removal of pollution in environmental water [48]. Figure 13 depicts the principle of preparation and characterization of such novel composites.

5 Waste Food Products-Based Bio Composites

5.1 Composite from Recycled Expired Milk for Multifunctional Applications

A hydrophobic composite was developed by denaturation and amination of expired cow milk. A durable, multifunctional and super hydrophobic composite can be prepared from the expired cow milk. It exhibits excellent water-repellence performance. A thermally denatured protein prepared from expired cow milk was heated and blended with nanoscale pigments of inorganic nature in order to develop the

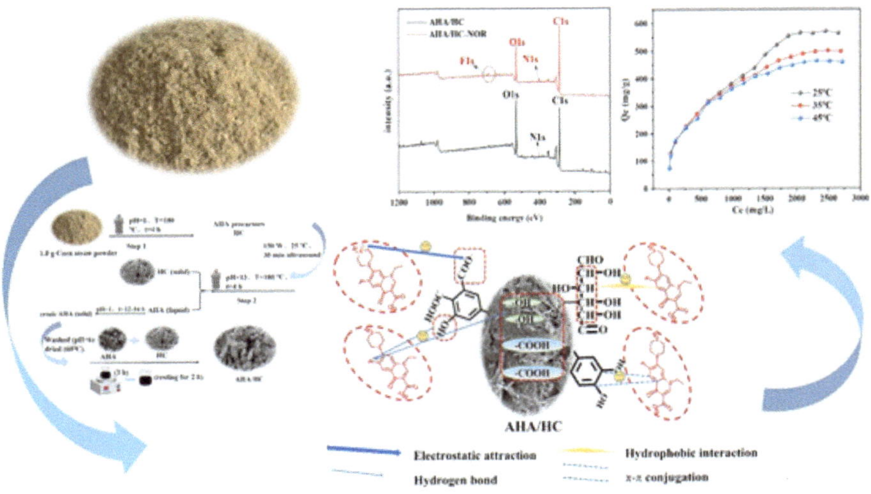

Fig. 13 Biomass composite for high-efficiency adsorption. Reproduced with permission from [48]

multifunctional biomass composite with excellent superhydrophobic performance [49–51]. Moreover, the fabrics coated with such material show an average contact angle of 155.7° and an average sliding angle of about 1.5°. Further, the expired cow milk can also be efficiently transformed into UV-photoluminescent, nonfluorinated and superhydrophobic composite by using synergistic and complex processes like denaturation and chemical coupling. The contact angle of such hydrophobic composites can be as high as 153.3° and the sliding angle can be about 2°. Depending on the desired functions of the biomass composite, the fields of application of such non-wettable and switchable materials can be decided. It is inferred that the expired cow milk-based composites showing multifunctional performances can expand their potential applications into the fields of protective materials, building materials for decorative purposes, fluorescent coatings, and protection of ice. The principle of recycling expired milk into a multifunctional composite is shown in Fig. 14.

5.2 *Recycling Fried Food Waste as a Binder for Developing Biodegradable Composites*

The collection of used cooking oil and its recycling is a well-established procedure. However, the food particles and debris discarded from the process of frying have not been given much attention. The fried food waste is found to be still having approximately 60% of cooking oils. Such residual oil can be extracted using some kind of squeezing process. The resulting dried solid waste as shown in Fig. 15 can be of significant importance for developing biomass composites using several other ingredients (e.g., powders derived from used cardboard, coffee husk and powder from rice

Fig. 14 Recycling of expired milk into composite. Reproduced with permission from [51]

straw etc.). The polysaccharides derived from solid fried food waste are excellent materials to be used as a binder in composites. The complex structures can be easily developed by using a heat-drying process at 90 °C for about 30 min followed by a treatment at 120 °C temperature for 30 min. This treatment depends on the shape of the composite. The developed composites exhibit a compressive strength as high as 2500 kPa and a thermal conductivity of about 0.089 W/(m·K). The shape of the developed composites can be maintained by using soaking with water. The compressive strengths might drop by 40% in high humidity conditions. These composites exhibit excellent performances especially in the areas of strong sorption affinity for toxic compounds, their possible biodegradability, lower levels of cytotoxicity, odor emission efficiency, and higher maize germination rates in soil beds. The performance and comparison of functions with other commercial materials like expanded polystyrene show that the use of recycled bio-based composites can have significant advantages in terms of ecological harmony, cost effectiveness and biodegradability at the end of life. Wastes generated from the fried food can potentially be a source of bulk bio material, which can help in fulfilling the objectives of circular economy at the same time providing alterative solutions to the society [52].

5.3 Phase Change Composites Developed from Lotus Shell-Based Biomass

Biomass from lotus shells is an abundantly available material with an excellent porous design and architecture. This biomass waste can be effectively used for producing

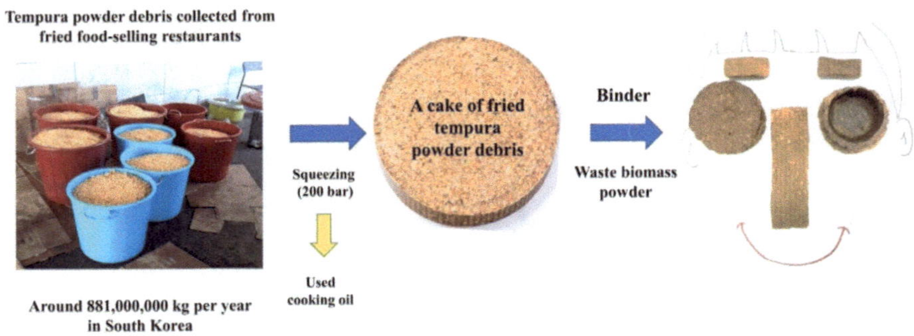

Fig. 15 Recycled waste from fried food industry in composite. Reproduced with permission from [52]

bio-based phase change composites (PCMs). Extensive research has been conducted to produce a series of shape-stable phase change composites (SSPCCs) based on n-docosane (ND) and biomass derived from waste lotus shells as shown in Fig. 16. This can improve the utilization of thermal energy, conversion of solar and photothermal energy, and in promoting environmental protection [53]. Carbonization and activation processes can play a very important role in the development and preparation of the pore structures. The carbonized lotus shells (CLS) and activated lotus shells (ALS) show much higher specific surface areas, in the range of 147.70 m^2 /g for CLS and 1652.35 m^2 /g for ALS respectively. The composite can also exhibit relatively high storage of latent heat reaching around 93.68 J/g, high level of prevention to leakage, and excellent stability at multiple thermal cycles. Such composite materials show a significantly high thermal conductivity of about 0.482 W/(m·K) with an increase of 189% compared to the base material. The phase change composites offer excellent chemical compatibility, relatively higher thermochemical stability, and excellent capability for overall thermal management. The waste lotus shell-derived bio-composites exhibit tremendous potential for thermal management, heat recovery applications, conversion of solar energy to other forms, and for regulating temperature in a given environment. It offers an alternative solution for environmentally sustainable composite material in application areas for energy storage and conversion. This is a promising approach towards sustainable composites for future generation solar cells and batteries where energy conservation and efficiency is of utmost importance [53].

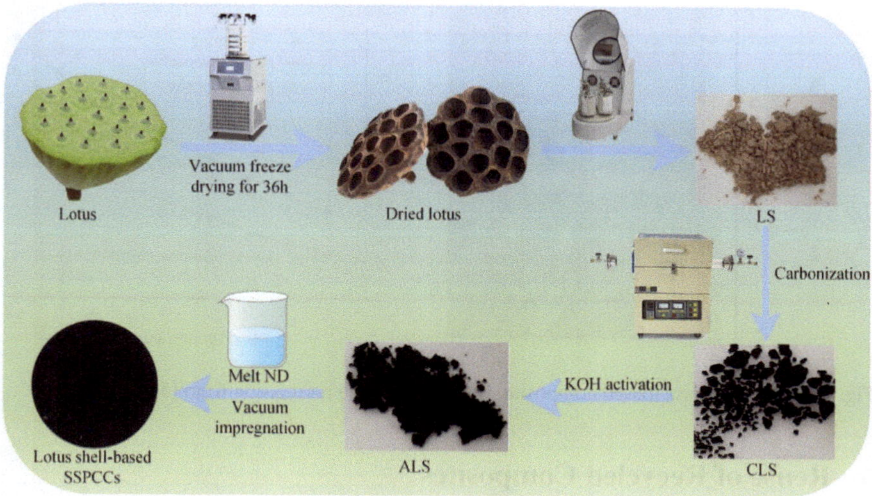

Fig. 16 Waste lotus shell-based composite. Reproduced with permission from [53]

5.4 *Functional Composites Using Biomass Derived from Algae Along with Graphene Oxide and Sodium Alginate*

Functional composites were developed by using graphene, sodium-alginate and Fe_3O_4 along with biomass derived from algae. Such composites can be used for the efficient removal of 2,4-D (2, 4-dichlorophenoxyacetic acid) in batch and continuous processes. The biomass from algae (Chlorella) needs to be pre-treated and then it can be used for the preparation of bio-composites as shown in Fig. 17. Fourier transform infrared spectroscopy (FTIR) and scanning electron microscopy (SEM) techniques are generally employed to determine and analyze the various functional groups and the morphology of the bio-composite materials. In the batch process, the parameters of adsorption need to be optimized. The biomass has to be pretreated with H_2SO_4. Its composites using other ingredients e.g., graphene/algae, Fe_3O_4/algae and sodium-alginate/algae show excellent biosorption capacities of 19.73, 20.43, 21.34 and 19.2 (mg/g), respectively for 2,4-D removal. The data for adsorption of 2,4-D follows the principle of Freundlich isotherm and model of pseudo-second-order kinetics. The rate of flow, height of bed and initial level of pesticide concentration have to be optimized in continuous mode. Desorption analysis also proves efficient recycling and reuse of effluent chemicals by using such bio composites. Research has proven that bio-composites using graphene, sodium-alginate and Fe_3O_4 along with biomass derived from algae are highly efficient adsorption materials for 2,4-D and can be effectively used for the removal and reuse of 2,4-D from the polluting industrial effluents [54].

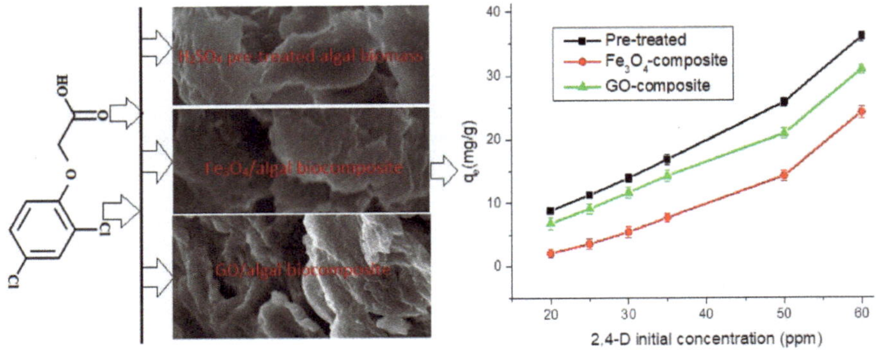

Fig. 17 Algal biomass -based composite. Reproduced with permission from [54]

6 Reuse of Recycled Composites

Ongoing research in the mechanical recycling of fiber-reinforced polymer composites aims to improve the processing of waste materials, optimize fiber recovery, and enhance the quality of recycled products. Several studies have contributed valuable insights into how fiber properties and recycled material quality can be optimized for further applications. It has been demonstrated that short, recycled fibers from composites can be effectively reused as reinforcing agents in thermoplastic-matrix composites via a simple regrinding process. The recycled fibers can be mixed with commercial fibers to create reinforced resin-based composites without adversely affecting mechanical performance. Reducing impurities and improving fiber surface cleanliness can further enhance the performance of these recycled products.

By carefully managing the process of separation and subsequent reformulation, the structural integrity of recycled composites can be maintained, and larger volumes of recyclate can be used. The novel and innovative techniques produce refined granules of recyclate materials, that can be used as an alternative for conventionally used fibrous reinforcement with minimal impact on functional properties. Adjusting the preparation procedure further enhances composite strength, making regrinding a viable solution for economically sustainable composite recycling. Key findings include that the specific energy demand for recycling depends on throughput, with lower demands at higher recycling rates. The energy model estimates specific cutting energy, providing valuable data for life cycle analysis (LCA). The separated materials are categorized based on size and quality. Recycled fibers are sorted by their length and mechanical properties, while resin powders and fillers are refined for reuse as secondary materials [36, 44].

The integration of recycled materials into 3D printing technologies is driven by the need for sustainable development. Thermoplastic materials like acrylonitrile butadiene styrene (ABS) and PLA, commonly used in fused filament/granular fabrication, are increasingly being derived from recycled biobased fibers. In addition to conventional materials, novel bio-based and cellulose-based composites have been introduced, further advancing the economic and environmental viability of 3D printing. The adoption of 3D printing with recycled materials can also significantly reduce emissions. For example, the recycling of plastics like PET, HDPE, and PP in AM processes has been shown to reduce emissions by up to 67%, 71%, and 71%, respectively. These figures underscore the environmental benefits of incorporating recycled materials into manufacturing processes, particularly when considering the short lifecycles of many components, prototypes, and molds. Furthermore, PLA-based composite materials prepared from natural fibers derived from bamboo, sisal, and wood cellulose are becoming increasingly popular in 3D printing. However, studies show that the mechanical properties of these composites may deteriorate due to the recycling process, primarily due to PLA degradation during the recycling process. This highlights the ongoing challenge of maintaining material integrity after recycling, which is crucial for ensuring the long-term viability of recycled composites in high-performance applications [49–53].

7 Concluding Remarks

Natural fibers are widely produced in many developed countries, creating employment opportunities and reducing environmental risks. Testing the degradation of natural fibers has become a common trend recently. The degradation rates of natural fibers have been investigated, revealing that the amount of fiber reinforcement in the composite improves degradation rates. The biodegradability and recycling of natural fibers is considered their most important and fascinating aspect in polymeric composite materials.

The design and development of advanced structural composites from biomaterials can eventually replace conventional man-made structures or even metals. This has led to innovative methods for developing recycled components. Designing lightweight, high-performance composite panels for heavy loads can significantly reduce consumption of fossil fuel and reduce emissions of carbon dioxide in the atmosphere.

Finally, the bio-based recycled composite materials may contribute towards well-being in society and can be used widely in several applications areas in the urban societies. Bio-based composites can be alternatives for various types of applications in the transportation industry, electric vehicles, interiors of automotive, structural elements in machine, aircrafts, satellites etc. Recent advancements in recycling of biocomposites are focused on optimizing fiber recovery and purity, particularly for composite applications. Technologies such as high-voltage fragmentation and precision shredding have shown promise in minimizing fiber damage and improving recyclate quality. These innovations ensure that recycled materials meet the performance standards for new accessories, reducing reliance on virgin materials and promoting a circular economy in the composite industry.

References

1. Wuzhou L, Kun Z, Liangang Z, Fujun X (2023) Molding and verification of dielectric constant of three-dimensional woven spacer composites. Composite Struct 304:116407. https://doi.org/10.1016/j.compstruct.2022.116407
2. Le Moigne N, Longerey M, Taulemesse JM, Bénézet JC, Bergeret A (2014) Study of the interface in natural fibers reinforced poly (lactic acid) biocomposites modified by optimized organosilane treatments. Ind Crops Products 52:481–494. https://doi.org/10.1016/j.indcrop.2013.11.022
3. Cordeiro N, Gouveia C, John MJ (2011) Investigation of surface properties of physico-chemically modified natural fibers using inverse gas chromatography. Ind Crops Products 33(1):108–115. https://doi.org/10.1016/j.indcrop.2010.09.008
4. Graupner N, Herrmann AS, Müssig J (2009) Natural and man-made cellulose fiber-reinforced poly (lactic acid)(PLA) composites: an overview about mechanical characteristics and application areas. Compos Part A Appl Sci Manuf 40(6–7):810–821. https://doi.org/10.1016/j.compositesa.2009.04.003
5. Khondker OA, Ishiaku US, Nakai A, Hamada H (2006) A novel processing technique for thermoplastic manufacturing of unidirectional composites reinforced with jute yarns. Compos Part A Appl Sci Manuf 37(12):2274–2284. https://doi.org/10.1016/j.compositesa.2005.12.030
6. Muthukumar K, Sabariraj RV, Dinesh Kumar S, Sathish T (2020) Investigation of thermal conductivity and thermal resistance analysis on different combination of natural fiber composites of Banana, Pineapple and Jute. Mater Today Proc 21:976–980. https://doi.org/10.1016/j.matpr.2019.09.140
7. Abral H, Atmajaya A, Mahardika M, Hafizulhaq F, Kadriadi, Handayani D, Sapuan SM, Ilyas RA (2020) Effect of ultrasonication duration of polyvinyl alcohol (PVA) gel on characterizations of PVA film. J Mater Res Technol 9(2):2477–2486. https://doi.org/10.1016/j.jmrt.2019.12.078
8. Aisyah HA, Paridah MT, Sapuan SM, Ilyas RA, Khalina A, Nurazzi NM, Lee SH, Lee CH (2021) A comprehensive review on advanced sustainable woven natural fiber polymer composites. Polymers 13(3):1–45. MDPI AG. https://doi.org/10.3390/polym13030471
9. Aly NM, Seddeq HS, Elnagar K, Hamouda T (2021) Acoustic and thermal performance of sustainable fiber reinforced thermoplastic composite panels for insulation in buildings. J Build Eng 40. https://doi.org/10.1016/j.jobe.2021.102747
10. Ashraf MA, Zwawi M, Mehran MT, Kanthasamy R, Bahadar A (2019) Jute based bio and hybrid composites and their applications. Fibers 7(9). MDPI Multidisciplinary Digital Publishing Institute. https://doi.org/10.3390/fib7090077
11. Begum S, Fawzia S, Hashmi MSJ (2020) Polymer matrix composite with natural and synthetic fibers. Adv Mater Process Technol 6(3):547–564. Taylor and Francis Ltd. https://doi.org/10.1080/2374068X.2020.1728645
12. Hamed E (n.d.) Load-carrying capacity of composite precast concrete sandwich panels with diagonal fiber-reinforced-polymer bar connectors. PCI J
13. Luhar S, Suntharalingam T, Navaratnam S, Luhar I, Thamboo J, Poologanathan K, Gatheeshgar P (2020) Sustainable and renewable bio-based natural fibers and its application for 3d printed concrete: a review. Sustainability (Switzerland) 12(24):1–25. MDPI. https://doi.org/10.3390/su122410485
14. Sabaruddin FA, Tahir PM, Sapuan SM, Ilyas RA, Lee SH, Abdan K, Mazlan N, Roseley ASM, Khalil Hps A (2021) The effects of unbleached and bleached nanocellulose on the thermal and flammability of polypropylene-reinforced kenaf core hybrid polymer bionanocomposites. Polymers 13(1):1–19. https://doi.org/10.3390/polym13010116
15. Sabet SA, Nazari S, Akhbari M, Kolahchi R (2019) Investigation of the bending behavior of 3D glass fabric-reinforced composite panels as slabs in buildings. Earthq Struct 16(3):369–373. https://doi.org/10.12989/eas.2019.16.3.369

16. Singh H, Inder J, Singh P, Singh S, Dhawan V, Tiwari SK (2018) A brief review of jute fiber and its composites. Mater Today Proc 5. www.sciencedirect.comwww.materialstoday.com/proceedings2214-7853
17. Syafiq R, Sapuan SM, Zuhri MYM, Ilyas RA, Nazrin A, Sherwani SFK, Khalina A (2020) Antimicrobial activities of starch-based biopolymers and biocomposites incorporated with plant essential oils: a review. Polymers 12(10): 1–26. MDPI AG. https://doi.org/10.3390/polym12102403
18. Syafri E, Sudirman, Mashadi, Yulianti E, Deswita, Asrofi M, Abral H, Sapuan SM, Ilyas RA, Fudholi A (2019) Effect of sonication time on the thermal stability, moisture absorption, and biodegradation of water hyacinth (Eichhornia crassipes) nanocellulose-filled bengkuang (Pachyrhizus erosus) starch biocomposites. J Mater Res Technol 8(6):6223–6231. https://doi.org/10.1016/j.jmrt.2019.10.016
19. Selver E (2019) Acoustic properties of hybrid glass/flax and glass/jute composites consisting of different stacking sequences. Tekstil ve Mühendis 26(113):42–51. https://doi.org/10.7216/1300759920192611305
20. Zheng L, Wang R, Yang X, Zhang K, Chen Y, Xu F (2024) Lightweight three-dimensional woven K evlar/glass hybrid composites with enhanced mechanical, dielectric, and thermal insulation properties. Polym Compo 45(7):6618–6628. https://doi.org/10.1002/pc.28222
21. Chen Y, Yuan F, Su Q, Yu C, Zhang K, Luo P et al (2020) A novel sound absorbing material comprising discarded luffa scraps and polyester fibers. J Cleaner Prod 245:118917. https://doi.org/10.1016/j.jclepro.2019.118917
22. Lee SH, Lum WC, Boon JG, Kristak L, Antov P, Pędzik M, Rogoziński T, Taghiyari HR, Lubis MAR, Fatriasari W, Yadav SM, Chotikhun A, Pizzi A (2022) Particleboard from agricultural biomass and recycled wood waste: a review. J Mater Res Technol 20:4630–4658. https://doi.org/10.1016/j.jmrt.2022.08.166
23. Zhu Z, Xiao P, Kang A, Kou C, Chen J (2023) Preparation of biomass composites with high performance and carbon sequestration from waste wood fibers. Constr Build Mater 404:133295. https://doi.org/10.1016/j.conbuildmat.2023.133295
24. Immerzeel P, Fiskari J (2023) Synergism of enzymes in chemical pulp bleaching from an industrial point of view: a critical review. Can J Chem Eng 101:312–321. https://doi.org/10.1002/CJCE.24374
25. Singh G, Sahu S, Bharti S, Arya SK (2024) Significance of enzymes for the recycling of wasted non-food biomass to value added products: a sustainable stewardship towards the cleaner environment, In: Process safety and environmental protection, Vol 190, Part A, pp 395–412, https://doi.org/10.1016/j.psep.2024.07.063
26. Khan S, Das P, Kasak P, AbdulQuadir M, Thaher M, Al Ghazal G, Faisal M, Hawari AH, Sayadi S, Al-Jabri H (2024) Production of sustainable thermoplastic composites from waste nitrogen fertilizer-grown marine filamentous cyanobacterium Geitlerinema sp. J Environ Manag 366:121931. https://doi.org/10.1016/j.jenvman.2024.121931
27. Gomes DA, de Novais Miranda EH, de Araújo Veloso MC, da Silva MG, Ferreira GC, Mendes LM, Júnior JB (2023) Production and characterization of recycled low-density polyethylene/amazon palm fiber composites. Indus Crops Products 201:116833. https://doi.org/10.1016/j.indcrop.2023.116833
28. Zhang F, Li J, Chen Y, Zhao S, Zhang Y, Wang R, He Y, Song P (2024) Combination of recycled cellulose with nano-thick flower layered double hydroxides: biomass water retention slow-release fertilizer for sustainable agriculture, colloids and surfaces a: physicochemical and engineering aspects 702. Part 2:135144. https://doi.org/10.1016/j.colsurfa.2024.135144
29. Bollakayala VL, Etakula N, Vuba KK, Uttaravalli AN, Ganta H, Dinda S, Gidla BR, Gadde L, Katiki G, Manda ST, Mutyapu S, Seelam NR (2023) Enhancement of wood-plastic composite properties in presence of recycled vehicular soot as a carbon source material: Sustainable management approach. Process Safety Environ Protection 174:286–297.https://doi.org/10.1016/j.psep.2023.04.013
30. Zhang J, Li X, Xu, Hang, Zhang W, Feng X, Yao Y, Ma Y, Su, Ling, Ren S, Li S (2023) Removal of Cd2+, Pb2+ and Ni2+ from water by adsorption onto magnetic composites prepared using

humic acid from waste biomass. J Clean Prod 411:137237. https://doi.org/10.1016/j.jclepro.2023.137237
31. Raj T, Chandrasekhar K, Naresh Kumar A, Kim S-H (2022) Lignocellulosic biomass as renewable feedstock for biodegradable and recyclable plastics production: a sustainable approach. Renew Sustain Energy Rev 158:112130. https://doi.org/10.1016/j.rser.2022.112130
32. Mandal A, Habib A, Paul B, Islam M, Ahamed B, Islam S, Sarker F (2024) Integration of recycled denim waste cotton fiber and jute fiber in thermoplastic bio composite applications. Results Mater 23:100611. https://doi.org/10.1016/j.rinma.2024.100611
33. Kim S, Kim H-J, Park JC (2009) Application of recycled paper sludge and biomass materials in manufacture of green composite pallet. Resour Conserv Recycling 53(12):674–679. https://doi.org/10.1016/j.resconrec.2009.04.021
34. Zheng R, Zhang X, Meng L, Wang X, Zhao L, Li H, Ren J (2024) Flexible, recyclable, and green fiber paper composites for HCHO degradation. Ind Crops Prod 218:119004. https://doi.org/10.1016/j.indcrop.2024.119004
35. Dinu R, Bejenari I, Volf I, Mija A (2024) Towards sustainable and recyclable plastic materials: Vanilin-based epoxy bio-composites with spruce bark powder and hydrochar, pioneering green chemistry solutions for shape memory applications. Ind Crops Prod 221:119363. https://doi.org/10.1016/j.indcrop.2024.119363
36. Huang X, Bo, Zhou, Sun G, Yang X, Wang Y, Zhang X (2023) Upcycling of plastic wastes and biomass to mechanically robust yet recyclable energy-harvesting materials. Nano Energy 116:108843. https://doi.org/10.1016/j.nanoen.2023.108843
37. Kshirsagar M, Kandasubramanian B (2024) Origami fabrication techniques for enhanced fiber reinforced composites: a review. Hybrid Adv 7:100274. https://doi.org/10.1016/j.hybadv.2024.100274
38. Lu C, Liu Y, Wang X, Yu J, Xu Y, Wang C, Wang J, Yong Q, Chu F (2022) Rigid, stretchable and full recyclable cellulose reinforced thermoset elastomer composites for photothermal conversion and Joule heating. Europ Polymer J 181:111710. https://doi.org/10.1016/j.eurpolymj.2022.111710
39. Xia Z, Feng Z, Niu Z, He J, Fan W, Bai C (2024) Biomass crosslinked fluorinated methacrylate elastomer for stress sensors with recyclability, reprocessability, remoldability, and hydrophobicity. Eur Polymer J 209:112888. https://doi.org/10.1016/j.eurpolymj.2024.112888
40. Lei Y, Zhang Y, Fu, Xiaowei, Wu, Bo, Lei J, Jiang L (2024) Biomass-derived dynamic polyurethanes: enhanced self-healing and recyclability through reversible cross-linking. Polymer 310:127461. https://doi.org/10.1016/j.polymer.2024.127461
41. Chen B, Liao M, Sun J, Shi S (2023) A novel biomass polyurethane-based composite coating with superior radiative cooling, anti-corrosion and recyclability for surface protection. Prog Org Coat 174:107250. https://doi.org/10.1016/j.porgcoat.2022.107250
42. Amiandamhen SO, Adamopoulos S, Adl-Zarrabi B, Yin H, Norén J (2021) Recycling sawmilling wood chips, biomass combustion residues, and tyre fibers into cement-bonded composites: Properties of composites and life cycle analysis. Constr Build Mater 297:123781. https://doi.org/10.1016/j.conbuildmat.2021.123781
43. Pedreño-Rojas MA, Porras-Amores C, Villoria-Sáez P, Morales-Conde MJ, Flores-Colen I (2024) Characterization and performance of building composites made from gypsum and woody-biomass ash waste: a product development and application study. Constr Build Mater 419:135435. https://doi.org/10.1016/j.conbuildmat.2024.135435
44. Lv Q, Zhu X, Zhou T, Tian L, Liu Y, Wang Y, Zhang C (2024) Multifunctional and recyclable aerogel/fiber building insulation composites with sandwich structure. Constr Build Mater 423:135902. https://doi.org/10.1016/j.conbuildmat.2024.135902
45. Escobar A, Sathicq Á, Pizzio L, Blanco M, Romanelli G (2015) Biomass valorization derivatives: clean esterification of 2-furoic acid using tungstophosphoric acid/zirconia composites as recyclable catalyst. Process Saf Environ Prot 98:176–186. https://doi.org/10.1016/j.psep.2015.07.008
46. Ramya V, Murugan D, Lajapathirai C, Meenatchisundaram S, Arumugam S (2022) A composite adsorbent of superparamagnetic nanoparticles with sludge biomass derived activated carbon

for the removal of chromium (VI). J Clean Prod 366:132853. https://doi.org/10.1016/j.jclepro.2022.132853
47. Woo D, Kang N, Park H, Myoung S, Lee G, Kim JK, Choi C (2024) Biomass-derived closed-loop recyclable chemically crosslinked polymer composites for green soft electronics. Chem Eng J 488:150818. https://doi.org/10.1016/j.cej.2024.150818
48. Chen W, Li X, Zhao P, Zhu X, Deng X (2024) Preparation of biomass artificial humic acid/hydrothermal carbon composite and its high-efficiency adsorption of norfloxacin. J Water Process Eng 66:105924. https://doi.org/10.1016/j.jwpe.2024.105924
49. Jing L, Wang Y, Li J, Lin X, Liu L, Chen Y, Liu H, Ying Z (2024) Innovative plasticization technique for talc-powder reinforced wheat-starch biomass composite plastics with enhanced mechanical strength. Int J Biol Macromolecules 269(Part 2):131894. https://doi.org/10.1016/j.ijbiomac.2024.131894
50. Sirajudheen P, Karthikeyan P, Ramkumar K, Nisheetha P, Meenakshi S (2021) Magnetic carbon-biomass from the seeds of Moringa oleifera@MnFe2O4 composite as an effective and recyclable adsorbent for the removal of organic pollutants from water. J Molecular Liquids 327:114829. https://doi.org/10.1016/j.molliq.2020.114829
51. Chen C, Gao C, Lin K, Zhang J, Zhang Z, Jing Y, Xiao Y, Shan G, Xie C (2024) Recycling of expired cow milk for constructing multifunctional biomass nonfluorinated chromatic paint with superhydrophobicity. Chem Eng J 492. https://doi.org/10.1016/j.cej.2024.152326
52. Oh MS, Yoon HY, Phong NT, Lee Y, Kang KH, Kim YM, Kim KS, Jeon J-R (2024) Recycling tempura powder debris derived from the fried food industry as a binder for 3-dimensional biodegradable composites: a novel circular economy alternative to low-performance plastics. Environ Technol Innov 36:103794. https://doi.org/10.1016/j.eti.2024.103794
53. Luo Y, Gao R, Pan Z, Xiao X, Zhang F, He X (2024) Biomass waste lotus shell-based shape-stabilized phase change composites with improved thermal conductivity for thermal management. Mater Today Commun 39:108769. https://doi.org/10.1016/j.mtcomm.2024.108769
54. Mustafa S, Bhatti HN, Maqbool M, Khan A, Alraih AM, Iqbal M (2024) Renewable functional composites of algal biomass with graphene oxide and Na-alginate for the adsorptive removal of 2,4-D herbicide. Sustain Chem Pharmacy 39:101577. https://doi.org/10.1016/j.scp.2024.101577

Animal Fibers Recycled Composite

Awais Ahmed Khan and Hafsa Jamshaid

Abstract Composite materials have been used in a wide range of applications since ancient times. In the past few years back the composites materials composed of synthetic materials/fibers such as carbon and glass fibers have gained much attention due to their extraordinary properties. However, in the present world, it is a crucial necessity to utilize natural fibers form renewable sources in composite development. Animal fibers offer extensive properties which are even comparable with commonly used synthetic fibers. This opens the chance to take advantage of animal fibers in composite manufacturing. In addition to this, huge amounts of animal fibers such as low graded wool, raw silk, feather, and hair fibers are discarded as waste. Exploiting such waste animal fibers offers the opportunity of significant resource recovery from biowaste. This chapter discusses some of the commonly used animal fibers such as wool, hair, silk, and feather fibers for composite production. Various manufacturing methods and the properties of composites materials are briefly described as well. This chapter also focused on the use of recycled animal fibers as potential material for sustainable natural fiber composites. Various methods for recycling of waste are studied and recycling techniques such as mechanical, chemical, and thermal techniques are investigated. It is evaluated in depth regarding positive and negative points of each technique for best selection. Comparison of properties of different composites and their application is investigated in detail and defined in terms of application.

1 Introduction

Composites are tailored materials that offer the combined properties of both reinforcement and matrix components. The potential benefits of fiber-reinforced composites include ease of processing, a wide range of properties, low cost, and a better strength-to-weight ratio. Composite materials exhibit the synergistic properties of both fibers' materials and matrix. Fibers reinforced composites have a broad spectrum

A. A. Khan (✉) · H. Jamshaid
School of Engineering and Technology, National Textile University, Faisalabad, Pakistan
e-mail: askhan181199@gmail.com

H. Jamshaid and U. Zubair (eds.), *Sustainable Recycling of Fiber Reinforced Polymer Composites*, SDGs and Textiles, https://doi.org/10.1007/978-981-96-9536-2_4

of applications in several fields and there is an ever-increasing demand in the market. The major share of fibers reinforced polymer composites is found in transportation, aerospace, wind energy, construction, electronic, and electrical industries. In year 2021, about 100 billion USD was the estimated production with a volume of 12.1 million tons. It is well known that in composites materials generally, carbon and glass fibers are used owing to their superior properties which enable their use in diverse range of applications. Various types of synthetic fibers are utilized in FRC. Synthetic fibers are used in composites for high end applications. Synthetic fibers are man-made so their properties can be engineered at the production stage which offer a greater advantage over natural fibers. Despite these benefits synthetic fibers are producing drastic impacts on the environment, the most severe being non-biodegradable waste. There is a prominent concern raised which is related to the usage of synthetic fibers composites as they cannot not be easily recycled so they are not eco-friendly and sustainable in a broader view. Natural fibers reinforced composites are a very good alternative and solution to this problem as natural fibers are environment-friendly, user friendly, and most importantly easily recycled. These composites are termed biocomposites or green composites. These biocomposites offer many advantages such as low cost, abundant availability of fibers, biodegradability, and ease in recycling. However, these natural fibers reinforced composites are readily used in the applications which do not require very high mechanical properties [1]. Demand for natural fibers-based biocomposite is increasing since past decade. There are wide variety of fibers available abundantly in nature, which can be utilized in the fibers reinforced composites. Natural fibers include plant, animal fibers, and mineral-based fibers. Plant fibers are also known as cellulosic fibers while animal fibers are known as protein fibers. Animal fibers are the second most important natural fibers after plant fibers. Animal fibers include various types of fibers such as silk, wool, Hair fibers, and feather fibers which are sourced from several animal sources. Various properties of natural fibers-reinforced composites depend verily on the characteristics of fibers such as fiber length, fiber strength and modulus, the orientation of fibers, and interaction between fibers and matrix. Mechanical properties of natural animal fibers are comparable with glass fibers in terms of specific modulus and strength. There are some challenges encountered in the course of broader usage of natural fibers such as their higher moisture content, lower resistance to fire, and mechanical impact. There are some concerns regarding the compatibility of natural fibers with the matrix type. On the other hand, there are some potential benefits of natural fibers which include their easy and abundant availability, low cost, biodegradability, low weight, and easy processing. Some latest studies revealed that the properties of natural fibers are comparable with the glass fibers reinforced composites. Within the context/domain of green composites, they are further classified as fully green composites and partially green composites. In fully green composites the resin material used is, also a renewable and sustainable material which can be easily degradable, whereas in partial green composites, the resin materials used are petrochemical-based [2].

Natural fibers including both plant and animal fibers are the integral element of the wide range of textile materials such as clothing, carpets, curtains, rugs, towels, and table cloths. Huge production and scraping of textile materials raise concerns about textile waste management. Various strategies for recycling of textile materials are introduced and still numerous studies are being conducted about the upcycling of textile materials. An innovative and alternative solution for waste management of textile products is the utilization of fibers in composite materials for various applications. In the composite industry, both virgin and recycled fibers can be effectively utilized. However, the essence of their usage has its own specific prospects. For example, using recycled fibers in the composite is beneficial in terms of both production and waste management. It provides an opportunity to reduce the quantity of landfilled waste produced due to textile products [3].

1.1 Commonly Used Animal Fiber in Composites

Fibers are mainly classified into two major categories, i.e., natural fibers and man-made fibers. Natural fibers are further categorized into three major categories: plant fibers, animal fibers, and mineral fibers. Plants fibers like sisal, jute, bamboo, and cotton fibers are widely used to produce natural fibers reinforced composites while in case of animal fibers their usage is comparatively very low. Here, in this chapter our primary discussion is on animal fibers and fibers-based composites. In next sections of the chapters commonly used animal fibers in composites, recycling of animal fibers and manufacturing techniques, and finally applications of animal fibers composites are discussed.

Fibers can be used in the matrix of composite in several ways either they can be directly used as fibers or fabrics composed of these fibers in form of woven, non-woven or knitted fabrics. The arrangement of fibers in the composites greatly influences the properties of composites. Beside this, natural fibers can also be used as fillers in the polymer composites to reduce their cost and also assist in making the composites more degradable material. Animal fibers are the second most important and abundant natural fibers after plants. As plant fibers contain a major portion of cellulose, animal fibers are mainly composed of proteins. Animal fibers are also referred as protein fibers. Animal fibers have high elongation and elastic recovery while lower stiffness. Animal fibers are mostly hydrophobic in nature even more than plant fibers. Furthermore, animal fibers are insulators of heat and highly sensitive to some specific Alkalis. Animal fiber composites are enticing the attention of researchers and producers to utilize them in a wide range of applications for low to moderate weight load bearing situations. These fibers are good alternatives to synthetic fibers. Animal fiber-based composites are used in different fields such as textile, biomedical applications, construction industry, and automobile industry. However, the properties and various characteristics of animal fibers-based composites have not yet been studied thoroughly so there is still need to further uncover their positive, sustainable aspects, and potentials [4]. Composites containing wool fibers as reinforcement exhibit results which are comparable with other natural plant fiber such as hemp fibers [5]. Animal fibers can be successfully used in composite materials as reinforcement for multi-axial applications. In composite production the

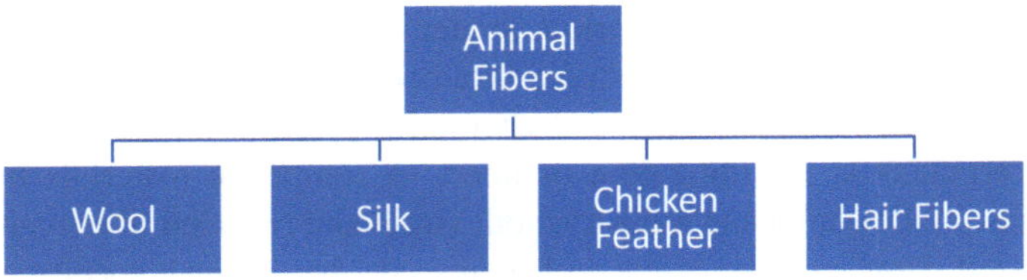

Fig. 1 Some commonly used animal fibers

commonly used animal fibers include wool, silk, feathers, Hair, etc. [6]. These fibers are explained in the coming sections (Fig. 1).

1.2 Silk

Silk is an outstanding animal fiber having potential to be utilized verily in animal fibers composites. Silk fiber is composed of proteins, and it is obtained from variety of sources such as silk worm, larvae of butterfly, spider among which mulberry silk obtained from silkworm is most common. Silk fibers obtained from silkworm is known as mulberry silk and silk fiber obtained from spider is called dragline silk. Both types of silk fibers are semi-crystalline in nature, mulberry silk has comparatively higher crystallinity than dragline silk. Mulberry silk has the crystallinity of about 38% to 66%. Silk fibers have spectacular attributes including high tensile strength, better elongation, and chemical resistance which make them feasible to use in composite development. Thermal degradation of mulberry silk fibers happens at 250 oC and dragline silk degrades above 170 oC [7]. Silk fiber is very delicate natural animal fiber with excellent mechanical properties, e.g., stiffness, strength. The durability of silk fibers has attracted the attention of researchers and producers to use it as a better alternative of the synthetic fibers in development of composites. It is reported in literature that the strength of dragline silk is even higher than polyamide fibers and comparable with some other synthetic fibers. Eco-friendliness with better properties are the key factors behind the appraisal of silk in composite industry [8]. Silk fiber is also a costly fiber due to its limited production. The large-scale usage of silk fibers in composites is not a feasible option for long term so waste silk fibers are the suitable option. In this regard, *Lalit Ranakoti* et al. developed hybrid biocomposite by adding industrial silk waste with jute fiber in epoxy composite. These composites were prepared using compression molding technique. It was observed that with the addition of silk waste fibers, the mechanical strength was improved and wear rate was reduced. Different animal fibers can be blended together and used in composite manufacturing to achieve the combined effect of both fibers. *Govindaraju* et al. used blends of silk and wool fibers with various blend ratios to produce composites with polypropylene matrix to evaluate the effect of fiber content on the impact properties, flexural, tensile, and dielectric properties. It was found that composites reinforced with only silk fibers exhibited better mechanical properties among all [9]. It is found in literature that about 11 million tons of silk waste is produced around the globe annually. Waste silk is obtained from three major sources which are wasted silk clothing comprise of silk, scraps produced during garment production, and during the silk formation process such as reeling waste. Waste silk fibers can be used in manufacturing of biocomposites as they will exhibit better properties [10].

1.3 Wool

Wool is another important natural animal fiber which is commonly used in textile and fiber reinforced composites. Wool fibers are obtained from several sources and named after their source like sheep wool, cashmere, alpaca, bison, angora, qiviut, etc. Wool fibers are composed of keratin protein. Wool fibers are hydrophilic in nature, they are mainly used for insulation purposes, wool fibers are prone to chemical action and environmental conditions. Wool is a coarser fiber as compared to other natural fibers including plant fibers. *Cashmere* wool is very rare and fine quality of wool obtained from cashmere goat which is quite rare and only available in highland habitat. Cashmere wool is much warmer, lighter in weight with greater repellency for water. *Khaliunaa* et al. conducted a study to evaluate the recycling efficiency of cashmere post-consumer waste including Pullovers, Jackets, and Scarves. It was concluded that the recycling of cashmere yarn was easy as it only involves the unraveling from the knitted structures however obtaining of fibers from retired garments were not quite possible [11].

In case of wool fibers, it is recommended that waste wool should be utilized in composites rather than virgin wool because of its high cost. It has been found in the literature that the felts or laminates composed of sheep wool fibers for the composite's materials have better heat insulation property than the polypropylene fibers in addition to this, wool fibers adhere with polymer matrix in a very good way. *Galan marin* et al. used wool fibers as reinforcement in natural composite soil to evaluate the impact of fibers on the mechanical properties of the soil present in wet climatic regions like Scotland. This experiment was performed in Scotland on soils with different compositions and it was found that the addition of wool fibers increased the stability by reducing the shrinkage rate and curing time along with increase in compressive strength. This approach of utilizing waste animal fibers in natural composites and improving the stability of soil can be helpful in the construction industry in wet climatic regions. Manufacturing of wool fibers reinforced composites is challenging in various aspects and require additional consideration. The super hydrophobic characteristics and inter-fibers cohesion may become problematic in the course of development of composites using compression molding method. Several wool fibers composites machining parameters like nozzle distance, jet pressure, and traverse rate of jet significantly affect the surface roughness of the composites. A need for an in-depth study of the wool fibers-based composites is required [12].

There are some wool fibers which are not suitable for textile applications so these fibers are regarded as waste fibers and can be efficiently utilized in composite manufacturing. In order to enhance the bio-degradability and eco-friendliness of these composites to the next level, biodegradable matrix material is used. Like Poly Lactic Acid (PLA). Textile waste comprising wool fibers were used to develop composite materials. *Zhong* et al. used waste wool fabrics, these wool fabrics were purified, decolorized and pulverized. The wool powder was then dissolved in the ionic solvent to extract wool protein keratin. Finally, PAN as matrix material was dissolved in N,N Dimethyl formamide along with wool keratin protein and ionic liquid and subjected to electrospinning to produce electro-spun membrane based on wool protein. This membrane possessed antibacterial properties up to 90% resistance against *E.coli* and 60% against *S.aureus* and feasible to use for the broad range of applications [13].

Using waste wool fibers in composite materials as reinforcement is very helpful in reducing the negative impact of wastage of wool fibers as well as composite materials. This is very beneficial in context of optimizing the cost of composite production by lowering the cost of raw material. It is reported that about one third of the total wool sheared from sheep is discarded as wastage based over its low-grade which is not suitable for textile applications. Presence of vegetable and mineral matter in wool fibers lowers the quality of wool fibers because the removal of this material is very difficult. Wool fibers can withstand the temperature up to 200 oC, there are hardly any natural fibers composites which can withstand such high temperature. Wool fibers composites can be compared with glass fibers composite; however, an extra point is the greater length of wool fibers which offers greater surface area and better adhesion with the resin which ultimately enhances the flexibility of the composites. In this way, the usage of synthetic materials like glass fibers can be decreased. Wool fibers can be coated with conductive materials to impart electrical conductivity into them. Using such treated wool fibers in composites have paved their way into an interesting application in suitable electrical component for example portable mini lights where lower and small-scale electrical power is used. This can minimize the usage of costly materials like copper, etc. Sheep wool fibers can be blended with other natural fibers in hybrid composites to improve certain properties. For example, *Jarin Tusnim* et al. reported that combining sheep wool fibers with treated jute fibers tends to enhance the mechanical properties and thermal stability of the hybrid composites [14]. In another work by *Dioisis Semitekolos* et al. he developed composites in which coarse sheep wool fibers were introduced in epoxy composites as toughening agent and this increased the thermal stability of sheep wool composites up to high degrees of temperature and the thermal conductivity coefficient of epoxy resin was raised up to 30%. Fibers composites comprising coarse sheep wool fibers tend to increase the composites' thermal stability and prolong the degradation of wool composites. This characteristic makes them feasible to use in the situations to cope up the fire emergency and provide greater response time [15]. Wool fibers are resistant to bacteria. Keratin protein whiskers aids in boosting the mechanical resistance of wool fibers. Limitation regarding the usage of wool fibers is that they absorb moisture verily. Wool fibers are hygroscopic in nature and can retain more moisture than any other natural fibers. The moisture absorption of fibers decreases the interfacial bonding between fibers and matrix, ultimately diminishing the composites' mechanical integrity. Increasing the percentage of wool fibers in the

Fig. 2 Wool fibers

composites results in lowering moisture absorption [16]. Wool being natural fibers have low compatibility for the variety of resin materials. So, in order to improve the adhesion of wool fibers with matrix materials various strategies are adopted. Silane coupling is one such very simple technology which performs the surface modification of wool fibers and improve their compatibility with the matrix material. In the study done by *Franciszek* et al. waste sheep wool fibers were obtained from the dairy industry and then surface modification was done using silane coupling agents. These silane functionalized wool fibers were then used with PLA matrix to produce composites. It was observed that the functionalized wool fibers have better computability than untreated which similarly improved the mechanical properties of the composites [17] (Fig. 2).

1.4 Chicken Feather

Chicken feather is another type of animal fiber which is composed of a fibrous protein namely β-Keratin which has very good mechanical and chemical resistance. Feather fibers are semi-crystalline in nature along with usual diameter ranges up to 5 μm. Chicken feathers are collected as a waste of the poultry industry which are usually disposed of or burnt without any potential usage. Poultry meat is consumed at very huge scale throughout the globe and its consumption is ever increasing (as shown in Fig. 3). In 2019, the estimated consumption of poultry meat was about 125milliion tons worldwide, in 2023, 103,418,000 metric tons of chicken meat was produced and it is predicted that by 2028 this consumption will cross over 140 million tons. The chicken feathers are produced as waste which are further utilized for the animal feed or disposed of as landfill. A very huge amount of chicken feathers is leftover as waste so these can be effectively utilized in the production of sustainable and green biocomposites. It is noteworthy that feather fibers have adequate mechanical properties such as tensile strength of feather fibers is in the range of 60 to 250 MPa and Young moduli in the range of 1 to 10 GPa. It can vary a little bit depending on the type of feather [18].

Composite manufacturing using chicken feather fibers is feasible and sustainable process. The keratin protein present in chicken feather fibers is hydrophobic in nature and it has strength comparable with the nylon fibers but the diameter of chicken feather fibers is comparatively smaller than wood fiber. The hydrophobic nature of chicken feather fibers ensures the better impregnation and fiber distribution with in the matrix [20]. Chicken feather fibers have outstanding characteristics of low density, acoustic absorption and good thermal insulation. Chicken fibers have greater acoustic absorption even higher than plant fibers [7]. Acoustic and thermal insulation properties of feathers fibers are credited to the hollow and honey comb cell morphology of feathers. The presence of such marvelous properties in feather fibers further broaden their spectrum of application in the composite industry. Moreover, composites materials containing feather fibers can be utilized for various end applications along with the potential biodegradability. Insulation in buildings is required to reduce energy consumption and the environmental impact of huge energy consumption. For this purpose, composite materials with better insulating

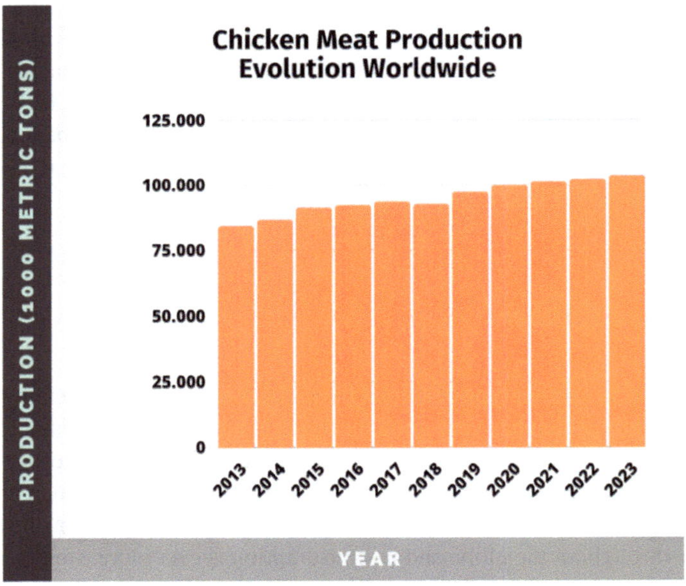

Fig. 3 Chicken meat production worldwide annually [19]

properties are introduced during construction. *Mustafa Khaleel* et al. fabricated rigid polyurethane foam using waste turkey feather fibers for insulation material for buildings. It was concluded that thermal insulation and acoustic resistance were increased with increasing the percentage of turkey feather fibers (TFF) and air permeability decrease. Mechanical properties of feathers fibers are very much dependent on the location of feather. The fibers taken from the closer parts have lower mechanical properties than the fibers taken from the rachis. The chicken feather fibers which are short and weak in nature and obtained from the closer part of chicken and other long chicken feather fibers are shown in Fig. 4. Still there is much work needed to further investigate the structure and properties of feather fibers [21]. Numerous researchers used chicken feather fibers in composites with various polymer resins and studied their properties. *Barone* et al. produced ecofriendly composites using keratin-based feather fibers with low-density polyethylene (LDPE) as resin. It is noteworthy that the density of keratin-based feathers fibers was found to be still less than LDPE and upon introduction of feather fibers the density of the composite reduced by 2 percent. On the other hand, the elastic modulus of the composites was also improved due to feather fibers [22]. *Bullions* et al. developed bio-composites using feather fibers with the help of compression molding technique in which multiple non-woven sheets of the fibers were compressed and polypropylene matrix were used. In this research work various types of natural fibers such as feather fibers, kraft pulp fibers, kenaf fibers were used and their properties were compared. It was found that the composites containing feather fibers exhibited properties comparable with other fibers even tension and bending properties of feather fibers were better than kenaf fibers reinforced composites [23]. In this domain, the duck down feather fibers is another important and widely available animal fiber that is still not used on large scale. Very little/less literature is reported about its use in composites manufacturing [20]. *Sanjay Kumar* et al. developed animal fibers reinforced composites using different animal fibers, i.e., goat hair fibers and bird feather fibers which include chicken feather fibers, duck, and turkey fibers. In this work, water absorption capacity was determined as a prominent problem in natural fibers reinforced composites. So, all the composites were compared and it was found that chicken feather fibers absorb least amount of water. Moreover, chicken feather fibers with epoxy matrix exhibited better mechanical properties such as tensile and flexural strength [24].

1.5 Animal Hair Fibers

Hair is the natural filament material mainly composed of keratin protein. Hair fibers are obtained from different types of animals, i.e., horse, cow, goat, camel, pig, etc. Animal hair fibers were used in the ancient times to strengthen the mud. The fine quality of the fibers which are obtained from certain species are utilized in the textile materials. A significant portion of other hair fibers are usually disposed of as waste materials. The need for proper end use and disposal of this waste is raised

Fig. 4 Waste chicken feather fibers

to avoid its impact on the environment. The utilization of hair fibers in the manufacturing of biodegradable and sustainable composites appeared as the ultimate solution. Waste animal hair fibers are also known as Bovine Hair waste (BHW) constituted of biological material known as keratinous material. These animal fibers are obtained in huge quality from the slaughterhouse as waste material which have no further use. These fibers are suitable to use in composite materials designed for low to moderate load bearing applications. Several studies reported the achievement of biodegradable polymer composite. Biodegradation aspect can be enhanced by increasing the content of BHW fibers in the composite's materials. There are studies reported regarding the enzymatic degradation of BHW fibers-based polymer composites which further favors their utilization in the development of biodegradable green composites materials. This further ensured the feasibility of their usage in development of green composites. Higher moisture absorption is the drawback of BHW fibers which limit the use of composites comprising BHW fibers. *Agbeboh* et al. conducted a study on the effect of volume fraction of BHW short fibers on the moisture absorption and abrasion properties of composites. In this work, hand layup technique was used to develop a composite with unsaturated polyester was used as matrix material and BHW short fibers were randomly oriented. It was observed that increasing the wt.% of fibers fraction result in increasing the moisture absorption percentage. It was found that upon 20wt% of fibers, the 0.2% moisture absorption reached up to 0.2% which is the threshold moisture absorption for natural fibers composites. It is a well-known fact that hair fibers have a hydrophobic nature because of the outer layer cuticle of the hair which prevents water absorption. In this case, still there is prominent moisture absorption by the BHW hair fibers in composites. The explanation to the matter is, the moisture absorption by hair fibers is due to the micro-pores present between cuticle scales and the greater affinity of the keratin protein toward the water. This implies that composites based on BHW fibers cannot be used in the water absorption-related application areas. It is noteworthy that BHW fibers result in improving the wear resistance of polyester composites due to the carbon content present in BHW fibers [25]. Many researchers studied the economical aspect of using waste hair fibers in construction of cheap and reliable composite materials. Cow hair fibers are the waste resulting from the leather industry. *Md Farhad Ali* et al. developed green composites using various percentages of cow hair fibers and unsaturated polyester resin. In this research, different percentages of cow hair fibers ranging from 2 to 20% weight were used. This was done to study their impact on the mechanical properties of the composites including tensile, flexural, and bending properties. It was concluded that the introduction of cow hair fibers improved the mechanical properties. Moreover, optimized mechanical properties were observed on the 5% weight addition of cow hair fibers [26]. Cow hair fibers are available in huge quantity as waste and byproduct. The meat of cow is edible while skin is normally used for the manufacturing of shoes, outfits, mats, etc. The hair fibers are usually disposed of by burning which also have negative impact on the environment so their application in the composite industry is very reasonable approach. *Isiaka O.oldaele* et al. conducted a study in which the cow hair fibers were used as reinforcement with HDPE matrix. Untreated cow hair fibers and chemically treated cow hair fibers with NaOH at 60 oC for one hour were used. The effect of fiber wt.% and chemical treatment on the flexural and tensile properties of the composites were studied. It was observed that increasing the fibers content tends to increase the flexural strength of composites. The chemical treatment of the fibers significantly affects the flexural and tensile properties by exhibiting better results as compared to untreated cow hair fibers. This is because the chemical treatment improves the bonding and adhesion of fibers with matrix material [27]. Horse hair is another example which are used for the production of economical and biodegradable composites. *Nitish* et al. developed biodegradable composites using horse hair. It was concluded that several properties of the composites were increased by using the horse hair such as flexural, tensile strength, and impact strength [28]. In the domain of hair fibers, goat hair fibers are the animal-based fibers which are obtained from the goat skin. The skin of goat is used in manufacturing of leather products and goat hairs are obtained as waste. These fibers are also biodegradable, easily available, soft, elastic, and low cost. The use of goat hair in composite materials offers a feasible solution for the disposal of waste obtained from tannery houses. *Gunti rajesh* et al. produced partially green composites using goat hair as reinforcement and polyester as resin and it was found that the mechanical properties of the composites increased up to a certain limit by increasing the fiber content and then a gradual decreases observed on further fiber addition [29].

Another animal hair fiber which is available in very huge quantity as a waste of food industry is pig hair which can be utilized properly as reinforcement in the composite materials. Human hair is another categorical animal fiber which is available in the form of huge waste and its disposal is a prominent issue as well. Human hair also has the potential to use in the composites. *Santos* et al. reported the use of human hair in the composites material for construction industry and study their compressive strength. Various weight percentages and lengths of human hair fibers were used to evaluate the effect of hair percent on the compressive strength. It was found that 0.72 inches hair length and 3.2% weight percent result in the optimum compressive strength [30] (Fig. 5).

Fig. 5 Wasted human hair fibers

1.6 Animal Processed Fibers from Manure

Animal-processed fibers are obtained from the manure such as dung of cow, sheep or other cattle. These fibers are actually plant material ingested by the animal and it is mainly composed of lingo-cellulosic material. Study conducted by *Vinayek Fasake* et al. suggested that the hemi cellulosic content ranges about 21–23% and cellulosic content about 36–38%. It is considered as one of the major sources of biogas and green energy. Cattle dung is easily and abundantly available material throughout the world. It is obtained in the form of farming waste. Numerous studies have been conducted to provide the proper solution and disposal of this waste. In the viewpoint of sustainability, utilization of such fibers from dung material in composites found to be a better and reliable solution. Cattle dung comprise of long as well as short fibers. Short fibers can be combined with other natural fibers in composites materials. Studies report that the dung fibers have the potential to be used in the composite materials for industrial application [31].

A research work done by the group of *Tom L Richard* assistant professor at Iowa state university reported that the manure of animals contains fibers as one of the major content after protein and water. These processed can be utilized for the composites materials for applications such as construction industry. It was reported that around 20 million tons processed animal fibers are generated annually. Some information is given in Table 1

The processed fibers replace the use of virgin animal fibers in the composite materials for construction industry. Protein content of manure is used as adhesive in the biocomposites [32].

Table 1 Animal processed fiber potential in the U.S

	Manure produced (dry tons/year)	Fiber content (as excreted)	Total fiber (dry tons/year)
Beef	12,100,00	44%	5,330,000
Dairy	23,200,000	42%	9,760,000
Swine	10,700,000	30%	5,300,000
Poultry	24,100,000	22%	3,200,000
Total	70,100,000		23.590,000

1.7 Various Methods for Recycling of Textile Waste

Textile waste can be upcycled to produce value-added products and counter the threat of excessive textile waste. However, recycling of textile wastes is somewhat a complex phenomenon because of insufficient management and blended or treated fibers is more complex. In order to obtain recycled animal fibers, discarded textile products are the major source. Waste animal fibers such as wool and silk can be obtained from the different stages of textile production and usage. Like the scrapes obtained during the garment manufacturing and textile clothing which are outdated are also available in the form of huge quantity of waste. So, there is need to recycled these products and get animal fibers. Sometimes animal fibers are present in the form of blends in textile materials. Composite industry offers huge capacity for the upcycling of textile waste by utilizing the textile waste in manufacturing of recycled composites. For example, the discarded waste fabric can be used as reinforcement in composite manufacturing. Secondly, textile waste in shoddy form can be used in composites and converting shoddy material into textile integrated structures such as yarns and fibers and then use them as reinforcement materials in composite manufacturing. Use of carbon fibers in the composite materials is on very large scale due to their demand in aerospace and automotive industry owing to the superior properties of carbon fibers. It is reported in the literature that about three kilo tons of carbon fibers waste is produced annually in U.K and USA within composite industry [33].

A brief overview of recycling technologies for textiles are discussed in this section.

1.7.1 Primary Recycling

The primary recycling is done for the recycling of textile materials comprising synthetic polymeric fibers such as Polyester and Nylon. In primary recycling mainly, such type of waste fibers is then re-melted and re-spun into the yarn form.

1.7.2 Secondary Recycling

In secondary recycling, waste material is used to make new product which is different from the product from which the waste is procured. In this recycling mechanical and chemical methods are used to recover the required material. Shredding is such example which is mechanical method to obtain fibers from textile materials and then these fibers are used in the production of door mats etc. the fibers obtained in the secondary recycling are used in manufacturing of composite materials.

1.7.3 Tertiary Recycling

In this recycling technique, the monomer units present in the products are recycled using chemical methods such as pyrolysis, hydrolysis or gasification. The monomers extracted in this method are reused to produce new products. For example, Worn-Technologies (Nottingham, England) developed closed-loop recycling system where they chemically extract the monomers from cellulose and polyester in fabrics in which they are blended from cellulose and polyester for production of yarn.

1.7.4 Quaternary Recycling

In this method of recycling, textile waste is used to generate heat. This method mainly focusses on the energy recovery rather than the reuse of material. So, in this context, textile materials are used as the source of energy production such as fuel and heat. In the same way, animal fibers like waste wool and feathers are burned to produce heat for various process. It reduces the landfill solid waste but it is not beneficial for recovery of animal fibers for composite. The type of polymer matrix material used with animal fibers significantly influence the properties of the composites.

1.8 Recycling Techniques for Recycling of Animal Fibers Composites

Recycling of composites is a bit complex phenomena which involves variety of sectors and stages for example engineering challenges, limited knowledge life cycle assessment, selection of material, and its development [34]. Generally broad scale recycling can be done for any type of synthetic fibers as they possess more integrity and can withstand the variety of forceful impacts. In case of animal fibers as they are natural fibers and have inherent structural and chemical integrity which is not enough to withstand any type of applied forces. In this regard, there are certain limitations as specific type of recycling method can be used for the specific type of fibers. An example is wool fibers which were previously only the closed loop recycling method was recommended in which only mechanical recycling of wool fibers via shredding, carding, etc., was done. So, in closed loop recycling strategy only 5% of the properties of the virgin wool were achievable. Like the mechanical recycling of wool fibers result in shortening the fiber length of wool fibers. There is much need to develop alternative and innovative methods in which the recycled fibers with better properties can be obtained [35]. Similarly, in the recent work reported by *Bingnan Mu* et al. wool fibers were regenerated protein fibers were produced using dissolution and wet spinning. In this method, the dye was also separated successfully and the protein fibers obtained were 90% like the virgin wool fibers. Still there is need to further study such methods to develop the recycling method for wool at large scale [36]. Fibers reinforced polymer composites are composed of matrix which can be natural or synthetic polymer. High demand, production, and consumption of composite materials produces huge amount of waste and landfill which have environmental concerns and recycling is an effective strategy to cope up the environmental concerns. In the process of recycling composites materials are converted into smaller entities which contain mixture of polymers and fillers and fibers. It is observed that commonly the finer materials in powder form mainly comprise of polymer and fillers while the coarser particles contain fiber materials. Usually, thermoset resins are used in fibers reinforced composites but with the recycling point of view thermoset resins composites are not easy to recycle. It is suggested that natural fibers composites can retain their thermo-mechanical properties until recycled 4 to 6 times and after that about 17% reduction occurs in their thermo-mechanical properties [37].

There are three main methods of recycling of composites which have different efficiency, working principle, potentials, and challenges. These three major methods are mechanical recycling, chemical recycling, and thermal recycling. Mechanical recycling, chemical recycling, and thermal recycling [38].

1.8.1 Mechanical Recycling

In mechanical recycling method composites materials are subjected to shredding, cutting, and grinding. These mechanical processes are done both using machine as well as manually, the shredder machine is shown in Fig. 6. This converts the material into smaller size which is easy to separate into fiber and polymer material. The extra small and fine particles produced during the recycling can be utilized for the variety of applications such as used as filler in composite manufacturing. In the slow speed mechanical recycling method, the materials size is reduced to 100 mm by the action of milling and cutting. The shredding, grinding, and crushing during mechanical recycling consumes energy. In mechanical recycling composites are first cut down into small pieces and then sort these fragments based on size. These small fragments help in separating the fiber material from resin. For the proper use of mechanical recycled composite material, there is an utmost need of size reduction of composite material and segregation of constituent material. Slow speed cutting and milling reduce the size up to 100 mm and further size reduction up to 10–50 mm can be done using high-speed milling usually done by hammer mill.

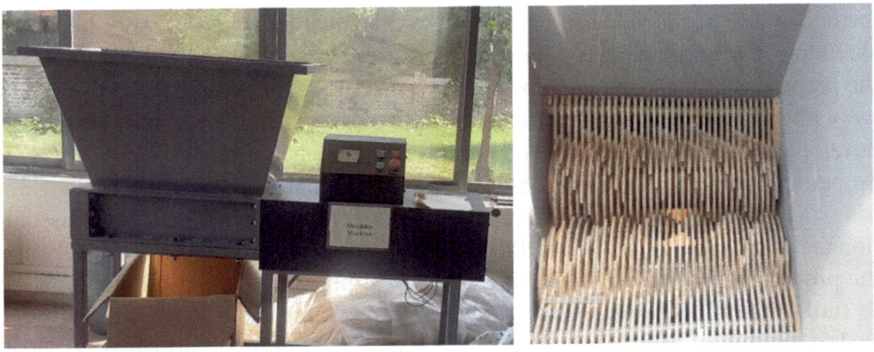

(a) Shredder Machine for mechanical Recycling of composites

(b) Inside view of Shredder Machine

Fig. 6 Shredder machine for mechanical recycling of composites

1.8.2 Thermal Recycling

Thermal recycling of composites is done by various recycling technologies such as combustion, pyrolysis and fluidized bed combustion. Thermal recycling of composites is done by providing external heat for the decomposition of composites into solid, liquid, and gas. However, there is challenge encounter in the course of thermal recycling regarding the disposal of residue produce in the form of ash. This raises the environmental concern of thermal recycling. There is problem regarding the use of this residue in virgin material as filler because it contains calcium oxide which was produced from calcium carbonate during thermal recycling which negatively affects the thickening process in virgin material production [39]. Thermal recycling is not suitable for natural fibers like animal fibers as mostly they are thermoset in nature and Beta sheet Nano crystallites fibers degrade upon heating. So, unlike synthetic fibers, thermal treatment such as melting and re-melting.

1.8.3 Chemical Recycling

Chemical recycling of natural fibers composites and animal fibers composites is the method in which the composite material is broken down into its constituents, i.e., matrix and fibers. The chemical recycling method involves the use of various solutions such as alkaline, acidic and enzymatic solutions. So based on these solutions chemical recycling of composites can be termed as acidic, alkaline or enzymatic. Beside this some other solvents like Toluene, acetone or Dichloromethane can be used to dissolve thermoplastic or thermoset resins. It is very important to note that there is immense care is taken in the selection of chemical in order to prevent the damage to the fibers. Acids solutions are commonly used to recycle the composites which contain thermoset resins or in which specific polymeric materials are used. It is well known that in case of animal fibers composites, acidic solutions are used very

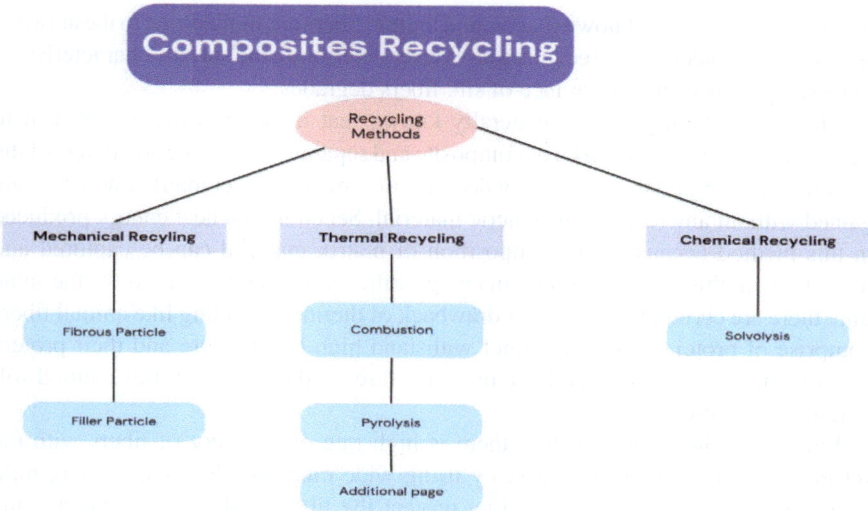

Fig. 7 Various composites recycling methods

cautiously because animal fibers are more prone to the acidic conditions and it can result in the degradation of protein structure of animal fibers as well (Fig. 7).

There is certain limitation in the recycling of fiber reinforced composites. Thermoset resins are mostly used in fiber reinforced polymer composites. Escalation of negative impact of waste materials and related concerns raise the demand to develop sustainable solutions.

1.9 Positive and Negative Effect of Specific Recycling Method/Pros and Cons of Recycling Methods

In the above section mainly three recycling methods are discussed. Each type of recycling method has its own positive and negative impacts on the properties of fibers. It is very important to choose the appropriate recycling method for specific type of fibers and their respective composites. In this section, the pros and cons of mechanical, chemical, and thermal recycling methods are explained.

Mechanical recycling is very economical and simple method as it is done at room temperature so it does not require any sort of elevated temperature conditions ultimately no extra energy is consumed. This is an ecofriendly method because it does not require the use of any harmful chemical. Moreover, the properties and chemical structure of fibers is preserved as no chemical is used. On the other hand, mechanical recycling method still possess negative impact on the properties of fibers such as it shortens the fiber length. Moreover, it alters the physical condition because the forceful impacts are exerted on the fibers. Like it can damage the crimp of wool fibers. Further it is well known that natural animal fibers are more prone to the abrasive stresses so the mechanical recycling obviously deteriorate the surface characteristics. For example, lustrous appearance of silk fibers degrades.

Thermal recycling as it is generally known that it involves the use of heat to decompose the matrix material of composite and separate fibers. The advantage of the thermal recycling method is, it fully decomposes the thermoset matrix and fibers are gained without any residual polymeric material. Secondly the heat energy produced in this method because of decomposition of matrix material can be captured and reused. So, in this way it becomes an energy-efficient recycling method. At the same time there are certain limitations or drawback of thermal recycling like animal fibers comprise of protein and they cannot withstand high temperature and their protein structure damaged. Additionally, to reach the threshold temperature this method still requires high amount of energy.

Finally, in chemical recycling there is high rate of recovery of fibers with the selective decomposition of matrix by using wide range of chemicals. Using mild chemicals and enzymes for recycling protect the fibers and avoid to damage the fibers. Contrarily much attention and care are needed to perform chemical recycling and choose optimum chemical solution because animal fibers are very sensitive to severe chemical conditions. Harsh and strong acidic and alkaline condition damage the surface and chemical attributes of animal fibers. Severe chemical action can affect the moisture properties and flexibility of the animal fibers for example the brittleness of wool fibers can increase significantly.

1.10 Manufacturing Method of Composites from Recycled Animal Fibers

Product design and manufacturing method is the first and foremost step in successful development of composites materials. This stage involves the careful selection of suitable natural fibers for the innovative product and the consideration of optimum processing method. The successful product design selection includes the consideration of economic, social, and environmental aspect of the material and processing technology and development method in order to minimize the environmental impact by controlling the carbon emission and wastage. Choice of manufacturing method of composite plays a crucial role in deciding the final properties of the composites. Several commonly used methods for the manufacturing of animal fibers composites.

1.10.1 Hand Lay-Up Method

Hand lay-up method is very simple as well as oldest method for manufacturing of composites. This method is used for the low volume or small batch production, this is cost effective and flexible method however the limitation is, the process is labor

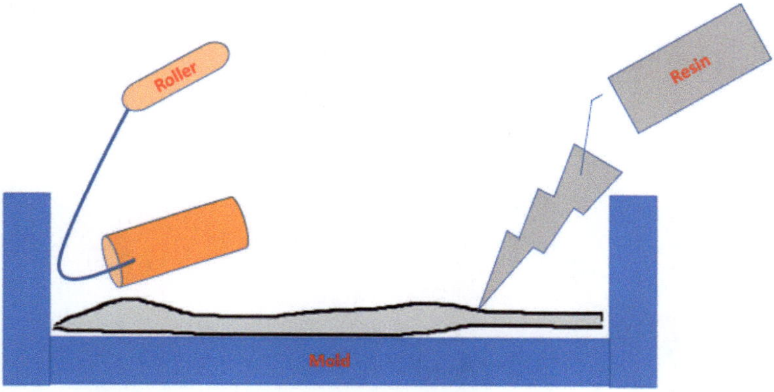

Fig. 8 Schematic diagram of hand lay-up technique

intensive. First of all, the suitable mold is selected keeping in view the design of final product. A thin layer of wax is initially coated on the mold to avoid the stickiness of composite with the mold surface. In this method, resin is manually applied using roller. Single or multiple layers of animal fibers are laid by hand with each layer being impregnated in the resin. Rolling is very necessary to ensure even distribution of resin and remove the voids and air bubbles ultimately improving the bonding between the fibers and matrix. Curing process is essential step in this method to keep uniformity. Curing is usually done at room temperature for animal fibers. This manufacturing method is readily utilized for the development of animal fibers-based composites. For example, *Sanjay kumar* et al. developed animal fibers reinforced composites with the help of hand lay-up method. In this two different types of animal fibers, i.e., animal hair (goat hair) and feather fibers (chicken feather and duck feathers) were used [24] (Fig. 8).

1.10.2 Spray up Method

Spray up method is the modification of conventional hand lay-up method. In spray up method, spray gun is used for the layering fibers and resin in the mold. The fibers are properly moistened and air is removed using roller similar to the hand lay-up method. This is comparatively faster method than the conventional lay-up technique. The advantage of this method is that it is a continuous process and wide variety of materials can be utilized as mold in the method. Moreover, the uniformity can be achieved by re-spray [40] (Fig. 9).

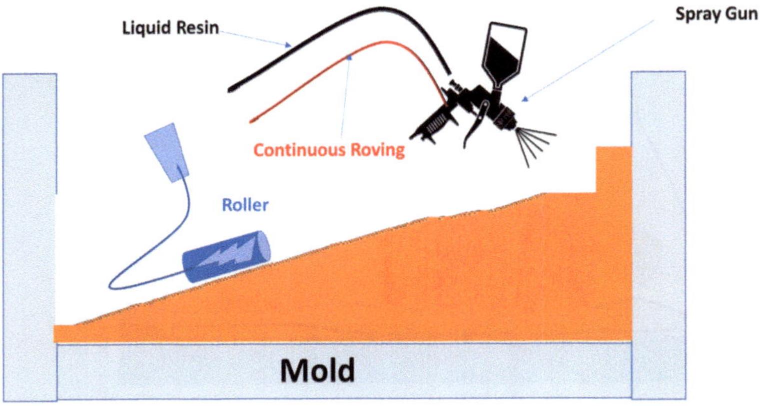

Fig. 9 Schematic illustration of spray up process

1.10.3 Vacuum Bag Molding Technique

This method is very feasible to manufacture natural fibers composites with good quality and uniformity. As the vacuum bagging ensure the removal of any void presents due to the fibers improper impregnation. *Sarah Paul* et al. developed hybrid composites using sheep wool and hemp fibers. In the reported work, the composites were prepared using vacuum bag molding technique. This method is very helpful in the development of sandwich structural composites. This method is also known as *Vacuum Assisted Resin Transfer Molding technique (VARTM)* and can be readily used for the development of composites materials using waste wool fibers [41]. In these composites woven and non-woven waste sheep wool fibers were used with epoxy resin. So, the arrangement of fibers in composites greatly influence the properties of the composites. It was found that the non-woven fibers laminate of waste wool fibers exhibited better interfacial bonding between the fibers and matrix while the woven laminates showed better hardness properties [42]. Vacuum Assisted Resin Transfer Molding Technique is the potential method for manufacturing of high-performance composites based on animal fibers. VARTM assure the better fiber compaction and resin penetration (Fig. 10).

Compression Molding is a technique reported in literature for the manufacturing of animal fibers reinforced composites. *Goncalves* et al. produced animal fibers reinforced polymer composites using Compression Molding technique. In these composites Portuguese Wool mats were used with thermoset polymer resin, namely Epoxy resin. Although, wool fibers demonstrated some level of difficulty in the course of impregnation in the polymer matrix. Likewise using of wool fibers without any treatment effects the behavior of fibers in composites. It was suggested that thermal treatment result in better behavior of wool fibers in the composites. This is due to the fact that thermal treatment tends to remove excessive humidity which is due to the presence of natural fat, ultimately enhance the adhesion of fibers with matrix [5]. *Isiaka O.oldaele* et al. produced animal fiber composites using cow hair fibers and

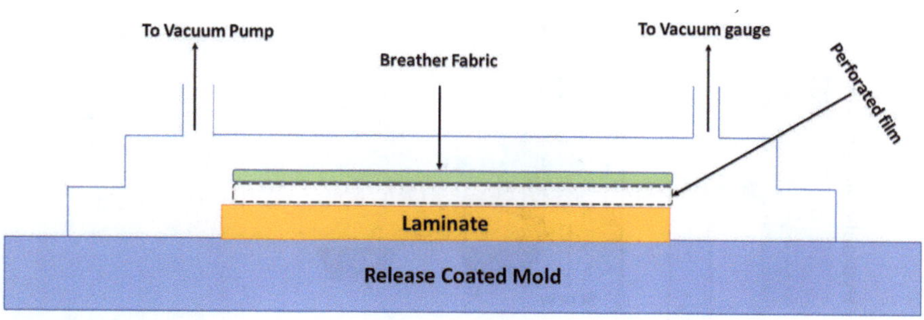

Fig. 10 Schematic diagram of vacuum bag molding technique

chicken feather fibers with HDPE as matrix material using compression molding technique. In another research work, wool and silk fibers-based hybrid composites were prepared using compression molding technique. In this method various percentage of both fibers with polypropylene matrix were used [43] (Fig. 11).

Melt Blending is another method to manufacture composites from recycled animal fibers, i.e., short wool fibers. *Lucia Conzatti* et al. produced composites using short wool fibers and polypropylene as matrix material using melt blending with internal batch mixture. It is worth mentioning that during manufacturing while using melt blending technique, it is very crucial to control the process conditions and parameters. Since the degradation of wool fibers occur at 250C and PP melts upon 165C beside this the friction occur between the wool fibers and interface materials tend to increase the temperature [44]. In a work reported by *ugur Soykan* et al. natural fibers composites were produced using feather fibers as reinforcement and thermoplastic

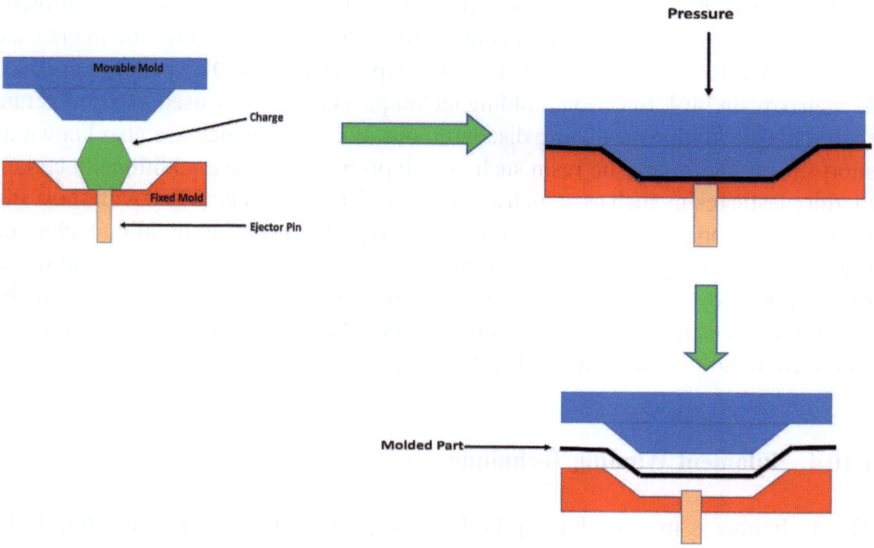

Fig. 11 Schematic illustration of compression molding technique

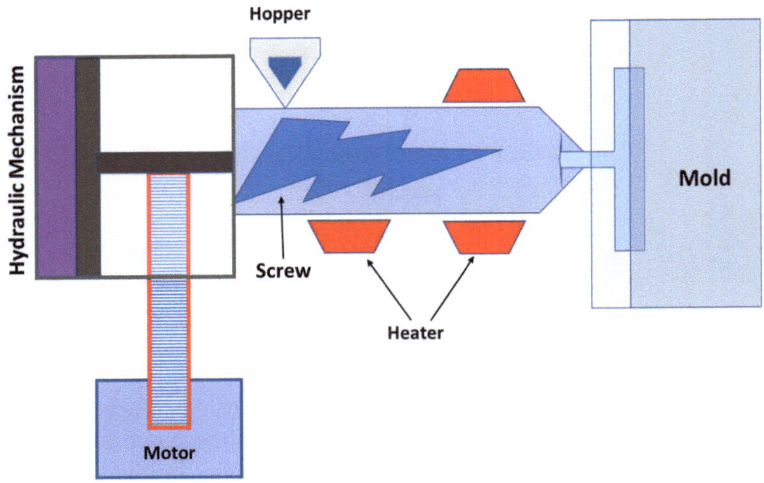

Fig. 12 Injection molding method setup

polyurethane (TPU) as resin. In this study the fabrication of composites was done by melt blending using micro injection molder and twin screw extruder. In this process, the temperature was kept constant at 190 oC for the time duration of 10 min. The temperature was kept constant in the mixing zone, hopper and die. The material was initially dried at 40 oC in the vacuum condition before the extrusion process and mixed with the help of twin screw extruder. The speed of screw was maintained at 100 rpm throughout the process [45].

Injection Molding Technique is another technique applied for the production of animal fibers reinforced composites. *Arinze* et al. produced animal fibers reinforced low-density polyethylene (LDPE) composites. In his research work, the fibers used were cow tail hair fibers, human hair, and sheep hair fibers. LDPE pellets were used as matrix/resin [46]. Injection molding technique is commonly used to manufacture thermoplastic composites using discontinuous silk fibers which are also known as short fibers. Thermoplastic resin such as polypropylene is used additionally certain thermoplastic resins such as, which are biodegradable, have been used with short silk fibers to develop composites. In these works, reported injection molding technique was recommended for the manufacturing of composites. It is noteworthy that in the work reported by *Mei po Ho* et al. significant improvement of 5% was observed in the flexural and tensile properties of short silk fibers PLA composites developed using an injection molding technique [47] (Fig. 12).

1.10.4 Filament Winding Technique

This technique is used to develop hollow composites. The filament winding technique for composites manufacturing was first introduced in 1960. It is crucial to optimize the parameters such as controlling the winding angle which should be kept between 20° and 87°. Silk is a natural filament which can be used in the development of filament wound composites. In the wide network of gas and petroleum, the pipelines in use are metal-based and prone to corrosion. So, the composite material pipes are alternative choice. This composite manufacturing method can result in achieving volume fraction and repetitive pattern. On the other hand, a very prominent limitation in this technique is that using this method the concave shapes cannot be produced [48]. Filament winding technique is used for manufacturing of composites with synthetic fibers such as glass fibers and carbon fibers. This method is not recommended commonly, for the manufacturing of natural fibers composites due the shorter fiber length and lower tensile strength of natural fibers. However, currently researchers are studying the feasibility of filament winding technique for natural fibers composites which have better strength and longer fiber length fibers like flax, sisal, silk, etc. Silk has higher crystalline nature as well as fiber length as it is also known as natural filament, silk can be used to produce filament wound composites [49]. Although there are not enough studies found in literature regarding the use of natural or animal fibers in the filament winding composites manufacturing methods. However, it is reported by *Vasi Uddin Siddiqui* et al. that wool fibers were used in filament winding process to manufacture composite tubes. In this work, wool fibers were continuously wound around the mandrel [50]. In a research work reported by *jiao Wen* et al. a similar setup of filament winding composite manufacturing was used in which the 3 K filaments of silk were passed through the pulleys and then impregnated into the liquid epoxy resin and then winding was done on the drum. The extra resin was removed and then the sheets were prepared as the impregnated filaments were removed from drum and it was found the thickness of these impregnated sheets were about 0.17 mm [51] (Fig. 13).

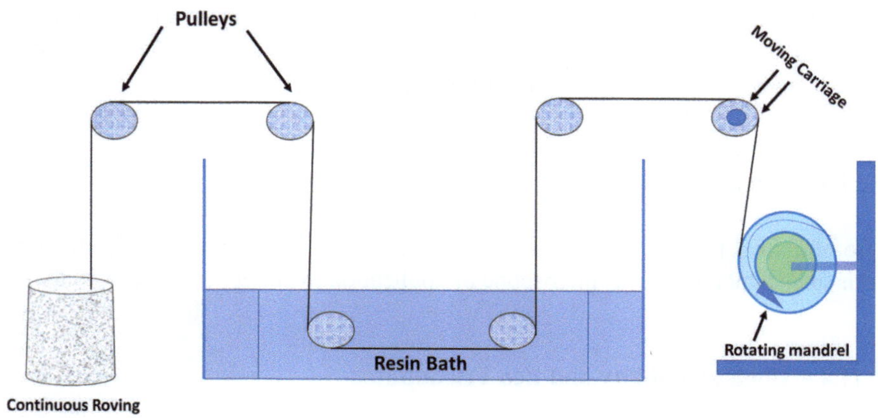

Fig. 13 Filament winding composite manufacturing setup

1.11 Properties of Animal Fibers Composites

Composites produced of recycled animal fibers such as recycled wool, wasted chicken feather fibers or raw silk fibers offer a wide range of properties which are comparable with the other type of fibers reinforced composites. Various properties of recycled animal fibers composites are briefly described as follows:

1.11.1 Mechanical Properties

Composites made with recycled wool fibers have sufficient flexibility and elasticity in order to use them in various applications where they are subjected to various loading and they don't break. In terms of strength and durability, it is observed that silk fibers reinforced composites are more durable and exhibit high tensile strength [52].

1.11.2 Thermal Properties

Composites manufactured with silk and wool fibers can withstand a broad range of thermal exposure as they have high resistance to heat. Moreover, they can be used in the areas where thermal insulation is desired. It is well known that the properties of fibers reinforced polymer composites are significantly affected by various factors, e.g., crystallinity of fibers that affects thermal stability, etc. In natural fibers, silk has highly crystalline nature which ultimately increases the thermal stability of silk-based composites [53]. *Bharath* et al. studied the mechanical, chemical, and thermal properties of sheep wool-reinforced fibers with vinyl ester and polyester composites. Better moisture and chemical absorption properties were observed when sheep wool fibers composites with vinyl-ester laminates [54].

1.11.3 Chemical Resistance

Use of animal fibers with appropriate resin materials also enhances their chemical resistance which increases their scope of application in variety of environments.

1.11.4 Biodegradability and Eco-Friendliness

Composite materials composed of recycled animal fibers are ecofriendly as they offer great advantage of waste utilization. Furthermore, as a natural resource they are biodegradable so there are no worries regarding their waste disposal. Composite manufacturing processes using recycled animal fibers materials have minimum carbon footprint and environmental footprint. Another beneficial aspect of animal fibers is that they are human friendly and it lacks in synthetic fibers like glass and carbon fibers which cause irritation when these are in contact with humans. Particularly, silk and wool fibers are hypo allergic and reduce the risk of allergies [52].

1.12 Comparison of Properties of Recycled Animal Fiber Composites with Virgin

Sometimes composites composed of recycled animal fibers with recycled synthetic fibers demonstrate even higher properties. It was observed that the composite material composed of 50% recycled wool fibers blended with 50% polyester fibers exhibited better acoustic insulation, moisture management, fire resistance than the composites comprising 100% recycled wool fibers and 100% recycled polyester fibers. In these composites, up to 70% biodegradability was also achieved by the end-of-life cycle (50 days) [55]. Recycled or waste animal fibers are not homogeneous in nature as well as they are usually short fibers so properties of composites deteriorate. Moreover, the animal fibers have great affinity for water molecules due to which they have low compatibility with the hydrophobic resins so there is an absolute need for chemical or thermal treatment to enhance their compatibility. *Palmer* et al. reported a study on the comparison between the mechanical properties of composites composed of recycled fibers and virgin fibers and found that using recycled fibers exhibited a prominent decrease in composites by about 18–30%. [56]. When recycled or waste fibers are used in composite materials as reinforcement, the contamination present in recycled fiber, which cannot be separated, acts as stressed breakage points which leads to the reduction in the mechanical properties of these fibers. On the other hand, use of virgin fibers result in better homogeneity throughout their arrangement in the composite structure which ultimately leads to the superior level properties of composite materials [57]. Using recycled fibers in composites reduces the tensile strength. However, it has also been observed that recycled fibers improve the strain strength because they are short and distributed randomly in the matrix.

1.13 Applications of Recycled Animal Fibers Composites

It has been found that natural fibers exhibit properties which are comparable with polymer and synthetic fibers and they possess additional advantages of being renewable and abundant in terms of existence, economical due to abundance in presence and biodegradable. Composites composed of animal fibers, both virgin and recycled, have a broader spectrum with reference to applications. Normally, it is suggested that the composites comprising recycled animal fibers should be limited from moderate to low load applications. Some application areas of animal fibers composites with respective examples are discussed below:

- Sustainable and energy efficient designs at minimum costs are the necessity of the time in the construction industry. Several research projects are focused on utilization of natural fibers composites in building industry. In this context, "Zero energy building (ZEB)" is trending approach in the European union keeping in view that building consumes energy and emits carbon with heating and cooling requirements with seasonal changes across the year. To minimize the ecological impact, light weight and ecofriendly composite materials based on waste animal fibers is the optimum choice. Building composed of natural fibers composite materials tend to regulate air and humidity inside the structure, additionally bringing a positive impact with the aromas of natural materials. *Jiri zach* et al. utilized sheep wool fibers to produce composites for construction of building designs and it was found that mechanical and thermal insulation properties were comparable and better in terms of performance than rock wool and mineral fibers. Additionally, it was confirmed that a reduced impact over human health was observed [58]. Sheep wool fibers-based composites were found to be very suitable for acoustic insulation applications [59]. *Rayed Alyousef* et al. developed green composites for building designs using waste sheep fibers and evaluated their performance. It was found that the use of waste wool fibers in concrete composites enhanced the sound and noise absorption capacity while reduced the compressive strength and modulus of elasticity values [60].
- The use of natural fiber composite materials in automotive industry results in decreasing the cost up to 20% and weight up to 30%. It is also found in literature that about 50% of the body part of air planes are composed of biocomposites comprising natural fibers. It was found that even Henry Ford emphasizes the use of green composites in order to strengthen the concept of recycle ability in the automotive industry [34].
- Wool fibers are used in the manufacturing of thermally insulated and biodegradable composites for packaging materials. Wool fibers, owing to the presence of keratin protein, have better flexibility, surface toughness, better aspect ratio and lower hydrophilic in nature than cellulosic fibers. Moreover, they have a low dielectric constant which makes them suitable for their use in the composite materials for electronic industry. Waste wool fibers and waste keratin protein which are obtained as the byproduct of the textile industry, owing to the low density, can be utilized efficiently for the development of lightweight composite materials for automotive applications [61].
- Animal fibers-based green composites are used in Bioengineering such as medical implants, i.e., dental implants, knee joints, templates for skin repair, artificial hip, heart valves, etc. It is found that natural fibers are very friendly to the human system [62].
- Natural fibers reinforced composites are widely used in the sport equipment's because of their lightweight they are preferred as they are easy to carry and move. For example, cricket bats, hockey stick, etc.

2 Conclusions

Animal fibers constitute the huge amount of waste which is produced around the globe every year. By using waste animal fibers in composite materials, it will be very beneficial both in terms of waste management as well as production of ecofriendly, biodegradable green composites materials. There are still certain limitations in the wide scale usage of animal fibers in composites. For example, the increase in the fibers content in the composites material resulting in the increase in water absorption property of the composites which is somehow problematic in various fields of applications. The interaction between the animal fibers and matrix is very weak which leads to void formation that ultimately impairs the certain properties. So, there is need of extra treatment of animal fibers prior to their application in composites. Selection of appropriate recycling and manufacturing methods is very important for the development of end application-oriented animal fibers composite. There is very limited data available in the literature in which the recycled animal fibers have been utilized for the composites. So, there is utmost need of in-depth study of potential of animal fibers in composite industry involving various recycling techniques which are feasible for obtaining good quality animal fibers. It is concluded from the available data related to animal fibers composites, that animal fibers have very high potential to be utilized in composite materials. Animal fibers-based composites are more durable than the plant fibers reinforced composites as plant fibers degrade at a higher rate than the animal fibers.

Animal fibers exhibit unique properties in their own aspects which appear to be potential for their utilization in composites for various applications. For example, the acoustic and thermal insulation properties offered by virgin as well as waste wool fibers are comparable with synthetic fibers. Likewise, silk fibers owing to the better crystal system offers better mechanical properties. This ensures their applications in different fields including biomedical applications, biodegradable packaging, automotive, electronics, and construction industries. Animal fibers constitute huge amounts of waste which is produced around the globe every year. By using waste animal fibers in composite materials, it will be very beneficial both in terms of waste management as well as production of ecofriendly, biodegradable green composites materials. There are still certain limitations in the wide scale usage of animal fibers in composites. For example, the increase in the fibers content in the composites material resulting in the increase in water absorption property of the composites which is somehow problematic in various fields of applications. The interaction between the animal fibers and matrix is very weak, which leads to void formation that ultimately impairs certain properties. So, there is a need for extra treatment of animal fibers prior to their application in composites. The selection of appropriate recycling and manufacturing methods is very important for the development of end application-oriented animal fibers composite. There is very limited data available in the literature in which the recycled animal fibers have been utilized for composites. So, there is utmost need of in-depth study of potential of animal fibers in composite industry involving various recycling techniques which are feasible for obtaining good quality animal fibers. Animal fibers-based composites are more durable than the plant fibers reinforced composites as plant fibers degrade at a higher rate than the animal fibers. Overall, animal fibers have very high potential to be utilized in composite materials for broad range of applications but this demand extensive studies.

References

1. Ramamoorthy SK, Skrifvars M, Persson A (2015) A review of natural fibers used in biocomposites: plant, animal and regenerated cellulose fibers. Polym Rev 55(1):107–162. https://doi.org/10.1080/15583724.2014.971124
2. Kumar S, Manna A, Dang R (2022) A review on applications of natural Fiber-Reinforced composites (NFRCs). Mater Today: Proc 50:1632–1636. https://doi.org/10.1016/j.matpr.2021.09.131
3. Suárez L, Castellano J, Díaz S, Tcharkhtchi A, Ortega Z (2021) Are natural-based composites sustainable? Polymers 13(14):1–17. https://doi.org/10.3390/polym13142326
4. Ghani A, Lee SH, Osman Al Edrus SS (2024) 1—An introduction to animal-based composites: possible animal-derived materials and the extraction process. In: Sapuan SM, Azhari CH, NMBT-PCD from Nurazzi AS (eds) Woodhead Publishing in Materials, Woodhead Publishing, pp 3–23. https://doi.org/10.1016/B978-0-443-22414-0.00001-6
5. Gonçalves S, Esteves JL Behaviour of wool fibres reinforced composite materials
6. Gholampour A, Ozbakkaloglu T (2020) A review of natural fiber composites: properties, modification and processing techniques, characterization, applications. J Mater Sci 55(3):829–892. https://doi.org/10.1007/s10853-019-03990-y
7. Ramesh M, Deepa C, Mavinkere Rangappa S, Siengchin S (2020) Biocomposites reinforced with animal and regenerated fibers. Hybrid Fiber Compos 197–215. https://doi.org/10.1002/9783527824571.ch11
8. Ranakoti L, Gupta MK, Rakesh PK (2019) Silk and silk-based composites: opportunities and challenges BT—Processing of green composites. In: Rakesh PK, Singh I (eds) Singapore: Springer Singapore, pp 91–106. https://doi.org/10.1007/978-981-13-6019-0_7
9. Rajkumar G (2022) Chapter 14 - Silk and wool hybrid fiber-reinforced polypropylene composites. In: Thomas S, Jose SBT-WFRPC (eds) The textile institute book series, Woodhead Publishing, pp 273–299. https://doi.org/10.1016/B978-0-12-824056-4.00016-9
10. Lu L et al (2022) Progress in recycling and valorization of waste silk. Sci Total Environ 830:154812. https://doi.org/10.1016/j.scitotenv.2022.154812
11. Davaadorj K (2016) Recycling of cashmere yarn and possible alternative fibers 2(Mic):1–30
12. Manivannan J, Rajesh S, Mayandi K, Siva I, Sankar I, Satyanarayana KG (2022) Chapter 17 - Abrasive water jet cutting and its optimization model for machining the sheep wool/polyester composites. In: Thomas S, Jose SBT-WFRPC (eds) The textile institute book series, Woodhead Publishing, pp 337–350. https://doi.org/10.1016/B978-0-12-824056-4.00014-5
13. Zhong X, Li R, Wang Z, Wang W, Yu D (2020) Eco-fabrication of antibacterial nanofibrous membrane with high moisture permeability from wasted wool fabrics. Waste Manage 102:404–411. https://doi.org/10.1016/j.wasman.2019.11.005
14. Tusnim J, Jenifar NS, Hasan M (2022) Effect of chemical treatment of jute fiber on thermo-mechanical properties of jute and sheep wool fiber reinforced hybrid polypropylene composites. J Thermoplast Compos Mater 35(11):1981–1993. https://doi.org/10.1177/0892705720944220
15. Semitekolos D, Pardou K, Georgiou P, Koutsouli P, Bizelis I, Zoumpoulakis L (2021) Investigation of mechanical and thermal insulating properties of wool fibres in epoxy composites. Polym Polym Compos 29(9):1412–1421. https://doi.org/10.1177/0967391120971387
16. Bharath KN, Pasha M, Nizamuddin BA (2016) Characterization of natural fiber (sheep wool)-reinforced polymer-matrix composites at different operating conditions. J Ind Text 45(5):730–751. https://doi.org/10.1177/1528083714540698

17. Pawlak F, Aldas M, Parres F, López-Martínez J, Arrieta MP (2020) Silane-functionalized sheep wool fibers from dairy industry waste for the development of plasticized pla composites with maleinized linseed oil for injection-molded parts. Polymers 12(11):1–22. https://doi.org/10.3390/polym12112523
18. Vilchez V, Dieckmann E, Tammelin T, Cheeseman C, Lee KY (2020) Upcycling poultry feathers with (Nano)cellulose: sustainable composites derived from nonwoven whole feather preforms. ACS Sustain Chem Eng 8(38):14263–14267. https://doi.org/10.1021/acssuschemeng.0c04163
19. www.chickenfans.com, "Poultry industry statistics"
20. Selvakumar K, Omkumar M (2020) Fabrication and characterization of animal fiber reinforced polymer composites (2):4538–4542
21. Khaleel M, Soykan U, Çetin S (2021) Influences of turkey feather fiber loading on significant characteristics of rigid polyurethane foam: thermal degradation, heat insulation, acoustic performance, air permeability and cellular structure. Constr Build Mater 308:125014. https://doi.org/10.1016/j.conbuildmat.2021.125014
22. Barone JR, Schmidt WF (2005) Polyethylene reinforced with keratin fibers obtained from chicken feathers. Compos Sci Technol 65(2):173–181. https://doi.org/10.1016/j.compscitech.2004.06.011
23. Bullions TA, Hoffman D, Gillespie RA, Price-O'Brien J, Loos AC (2006) Contributions of feather fibers and various cellulose fibers to the mechanical properties of polypropylene matrix composites. Compos Sci Technol 66(1):102–114. https://doi.org/10.1016/j.compscitech.2005.03.017
24. Sm S, Jayakrishna S (2023) Effect of fiber composition on physical and mechanical properties of turkey, duck, chicken feather and goat hair fiber reinforced polymer composites. EDXJL Int J Innov Adv Res 01:22–27. https://doi.org/10.59599/EDXJL-IJIAR.2022.1104
25. Agbeboh NI, OlaJide JL, Oladele IO, Babarinsa SO (2019) Kinetics of moisture sorption and improved tribological performance of keratinous fiber-reinforced ortho-phthalic polyester biocomposites. J Nat Fibers 16(5):744–754. https://doi.org/10.1080/15440478.2018.1434849
26. Ali MF, Hossain MS, Ahmed S, Sarwaruddin Chowdhury AM (2021) Fabrication and characterization of eco-friendly composite materials from natural animal fibers. Heliyon 7(5):e06954. https://doi.org/10.1016/j.heliyon.2021.e06954
27. Oladele IO, Olajide JL, Ogunbadejo AS (2015) Open access the influence of chemical treatment on the mechanical behaviour of animal fibre-reinforced high density polyethylene composites American journal of engineering research (AJER). Am J Eng Res (AJER) 4(2):19–26
28. Kumar N, Singh A, Debnath K, Ranjan R (2020) Mechanical characterization of animal fibre-based composites. Indian J Fibre Text Res 45(3):293–297. https://doi.org/10.56042/ijftr.v45i3.26218
29. Krishnasamy P. Evaluating tensile properties of animal and hybrid fiber reinforced polyester composites evaluating tensile properties of animal and hybrid fiber reinforced polyester composites. https://doi.org/10.1088/1757-899X/390/1/012011
30. Santos FD, et al (2020) Optimization of the compressive strength of hair fiber reinforced concrete using central composite design. In: IOP conference series: materials science and engineering, vol 771, no 1. https://doi.org/10.1088/1757-899X/771/1/012059
31. Fasake V, Dashora K (2024) Characterization of raw and anaerobic digested cattle dung fibers: a sustainable source of non-wood material. Biomass Convers Biorefinery 14(2):1949–1958. https://doi.org/10.1007/s13399-022-02487-0
32. TLR (I. S. University) (2015) Thinking outside the box: building materials and other products from animal processed fiber. AgEcon Search, p 18
33. Kamble Z, Behera BK (2021) Upcycling textile wastes: challenges and innovations. Text Prog 53(2):65–122. https://doi.org/10.1080/00405167.2021.1986965
34. Cestari SP, Silva Freitas DF, Rodrigues DC, Mendes LC (2018) Recycling processes and issues in natural fiber-reinforced polymer composites. Green Compos Automot Appl 285–299, 2018, https://doi.org/10.1016/B978-0-08-102177-4.00012-4

35. Celep G, Tetik GD, Yilmaz F (2022) Limitations of textile recycling: the reason behind the development of alternative sustainable fibers. In: Ibrahim H (ed) Rijeka: IntechOpen, p Ch. 5. https://doi.org/10.5772/intechopen.105118
36. Mu B, Yu X, Yang Y (2023) Sustainable and green process for recycling waste wool textiles into high-quality protein fibers on a pilot scale. Resour Conserv Recycl 198:107190. https://doi.org/10.1016/j.resconrec.2023.107190
37. Zhao X et al (2021) Recycling of natural fiber composites: Challenges and opportunities. Resour Conserv & Recycl X 177:105962. https://doi.org/10.1016/j.resconrec.2021.105962
38. Fuentes Molina N, Fragozo Brito Y, Polo Benavides JM (2021) Recycling of residual polymers reinforced with natural fibers as a sustainable alternative: a review. https://doi.org/10.3390/polym13213612
39. Asmatulu E, Twomey J, Overcash M (2014) Recycling of fiber-reinforced composites and direct structural composite recycling concept. J Compos Mater 48(5):593–608. https://doi.org/10.1177/0021998313476325
40. David AC, Naidu AL. A review on manufacturing techniques and chemical properties of natural composite materials [Online]. www.ijrdt.org
41. Ayyampilli SP, Suresh R, Hegde MG (2024) Effect of hemp and sheep wool reinforcements on mechanical and acoustic properties in natural fibre polymer composite using vacuum bag moulding process. J Inst Eng (India): Ser D. https://doi.org/10.1007/s40033-024-00807-4
42. Kishor Sharma Y, Meena A, Sahu M, Dalai A (2023) Experimental investigation on mechanical and thermal characteristics of waste sheep wool fiber-filled epoxy composites. Mater Today: Proc. https://doi.org/10.1016/j.matpr.2023.01.157
43. Rajkumar G (2022) Silk and wool hybrid fiber-reinforced polypropylene composites. In: Wool fiber reinforced polymer composites, Elsevier, pp 273–299. https://doi.org/10.1016/B978-0-12-824056-4.00016-9
44. Conzatti L et al (2013) Composites based on polypropylene and short wool fibres. Compos A Appl Sci Manuf 47:165–171. https://doi.org/10.1016/j.compositesa.2013.01.002
45. Soykan U (2022) Development of turkey feather fiber-filled thermoplastic polyurethane composites: thermal, mechanical, water-uptake, and morphological characterizations. J Compos Mater 56(2):339–355. https://doi.org/10.1177/00219983211056137
46. Arinze RU, Oramah E, Chukwuma EC, Okoye NH, Chris-Okafor PU (2022) Mechanical impact evaluation of natural fibres with LDPE plastic composites: waste management in perspective. Curr Res Green Sustain Chem 5:100344. https://doi.org/10.1016/j.crgsc.2022.100344
47. Ho M, Wang H, Lau K, Leng J (2013) Effect of silk fiber to the mechanical and thermal properties of its biodegradable composites. J Appl Polym Sci 127(4):2389–2396. https://doi.org/10.1002/app.37539
48. Colombo C, Vergani L (2018) Optimization of filament winding parameters for the design of a composite pipe. Compos B Eng 148:207–216. https://doi.org/10.1016/j.compositesb.2018.04.056
49. Mindermann P, Gil Pérez M, Knippers J, Gresser GT (2022) Investigation of the fabrication suitability, structural performance, and sustainability of natural fibers in coreless filament winding. https://doi.org/10.3390/ma15093260
50. Siddiqui VU, Sapuan SM, Segar JR, Seathuraman D, Murugasu L (2024) Wool fiber–reinforced polymer composites. In: Polymer composites derived from animal sources, Elsevier, pp 25–45. https://doi.org/10.1016/B978-0-443-22414-0.00002-8
51. Wen J, Zeng Y, Wu C, Guan J, Guo H (2020) Silk lattice structures from unidirectional silk fiber–reinforced composites for breaking energy absorption. Adv Eng Mater 22(3). https://doi.org/10.1002/adem.201900921
52. Mann GS et al (2023) Green composites based on animal fiber and their applications for a sustainable future. Polymers (Basel) 15(3):601. https://doi.org/10.3390/polym15030601
53. Thyavihalli Girijappa YG, Mavinkere Rangappa S, Parameswaranpillai J, Siengchin S (2019) Natural fibers as sustainable and renewable resource for development of eco-friendly composites: a comprehensive review. Front Mater 6(September):1–14. https://doi.org/10.3389/fmats.2019.00226

54. Bharath KN, Madhu P, Yashas Gowda TG, Verma A, Sanjay MR, Siengchin S (2021) Mechanical and chemical properties evaluation of sheep wool fiber–reinforced vinylester and polyester composites. Mater Perform Charact 10(1):99–109. https://doi.org/10.1520/MPC20200036
55. Patnaik A, Mvubu M, Muniyasamy S, Botha A, Anandjiwala RD (2015) Thermal and sound insulation materials from waste wool and recycled polyester fibers and their biodegradation studies. Energy Build 92:161–169. https://doi.org/10.1016/j.enbuild.2015.01.056
56. Palmer J (2009) Mechanical recycling of automotive composites for use as reinforcement in thermoset composites
57. Rusdianti, "utilizing sheep fibers for composites," 2024
58. Zach J, Korjenic A, Petránek V, Hroudová J, Bednar T (2012) Performance evaluation and research of alternative thermal insulations based on sheep wool. Energy Build 49:246–253. https://doi.org/10.1016/j.enbuild.2012.02.014
59. Abdellah MY, Sadek MG, Alharthi H, Abdel-Jaber GT (2023) Mechanical, thermal, and acoustic properties of natural fibre-reinforced polyester. Proc Inst Mech Eng, Part E: J Process Mech Eng 238(3):1436–1448. https://doi.org/10.1177/09544089231157638
60. Alyousef R (2022) Enhanced acoustic properties of concrete composites comprising modified waste sheep wool fibers. J Build Eng 56:104815. https://doi.org/10.1016/j.jobe.2022.104815
61. Kumartasli S, Avinc O (2024) 3—Biocomposites derived from animal fibers and other minerals. In: Woodhead publishing series in composites science and engineering. In: Ramnath RA, Sanjay MR, Gorbatyuk SM, Fischer S, SBTB-B-BF, Siengchin RR (eds) Woodhead Publishing, pp 55–78. https://doi.org/10.1016/B978-0-323-97282-6.00012-8
62. Raton B, New LF (2010) Group, Nano- and Biocomposites (1st ed). CRC Press (2009)

Recycled Natural Mineral Fiber-Reinforced Composites (MFRC) and Their Applications

Muhammad Shoaib and Usman Zubair

Abstract Issues in disposal and recycling have elevated concerns about the environmental impact of fiber-reinforced polymer composites (FRPCs), especially mineral fiber-reinforced composites (MFRCs), which are in high demand in industries like electronics, construction, aerospace, and automobiles. The strong adhesion between mineral fibers and polymer matrices renders conventional recycling methods—mechanical, thermal, and chemical labor-intensive and technically complex, despite the materials' well-known mechanical strength, thermal stability, and durability. Thermal approaches need a lot of energy and may damage fibers, chemical methods are expensive and difficult to scale, and mechanical operations typically destroy fibers, lowering material quality. Disassembly, shipping, and pre-treatment are all logistical hurdles that increase the financial cost and prevent broad adoption. Improved fiber recovery, material integrity, and economic viability may be achieved via new recycling processes, which are necessary to overcome these obstacles. Emerging technologies such as fluidized bed recycling and microwave-assisted systems have the potential to maximize efficiency while decreasing energy consumption. Hybrid systems that integrate mechanical, thermal, and chemical processes also show promise. Preserving the mechanical performance of recycled fibers, especially those susceptible to deterioration, and attending to the unique requirements of various mineral fiber types are of utmost importance. These developments, when combined with life cycle costing (LCC) and life cycle assessment (LCA) techniques, may provide light on the ecological and financial viability of different recycling strategies. To aid in the shift to a circular economy and the proper handling of these high-performance materials, these endeavors will be critical in creating long-term, scalable, and affordable solutions.

Abbreviations

BF Basalt fiber

M. Shoaib (✉) · U. Zubair
School of Engineering and Technology, National Textile University, Faisalabad, Pakistan
e-mail: ranashoaib.shoaib@gmail.com

H. Jamshaid and U. Zubair (eds.), *Sustainable Recycling of Fiber Reinforced Polymer Composites*, SDGs and Textiles, https://doi.org/10.1007/978-981-96-9536-2_5

CAPEX	Capital expenditure
CBA	Cost benefit analysis
CFRPs	Carbon fiber reinforced polymer composites
EOL	End of life
FRPCs	Fiber-reinforced polymer composites
GFRPs	Glass fiber reinforced polymer composites
GWP	Global warming potential
LCA	Life cycle assessment
LCC	Life cycle costing
MFPCs	Mineral fiber reinforced composites
NFRPCs	Natural fiber reinforced polymer composites
OPEX	Operational expenditure
PLA	Polylactic acid
rCF	Recycled carbon fiber
SCW	Super critical water
UCF	Cost per recovered fibers
UCW	Cost per mass of waste
VOC	Volatile organic compounds

1 Introduction

The worldwide market for MFRCs is predicted to have substantial development from its 2019 valuation of USD 228 billion, increasing to an anticipated USD 375 billion by 2026, at a compound annual growth rate (CAGR) of 7.3% [1, 2]. The production of these materials in 2021 was valued at USD 100 billion and accounted for 12.1 million tons of volume. They are fundamental to more than fifteen sectors due to their high mechanical strength, thermal stability, and durability. A few of the fastest-growing industries include building and transportation as well as wind power, aircraft, electricity, and electronics [3, 4]. Nevertheless, the difficulties in managing the MFRCs' lifetime are growing in tandem with their demand, particularly in view of the tightening environmental restrictions and waste reduction initiatives. For instance, by 2035, no more than 10% of municipal garbage may be sent to landfills, as per EU Directive 2018/850. Sustainable waste management procedures for MFRCs have become more important as a result. One solution is to adopt a circular economy strategy, which involves reusing and recycling materials via closed-loop systems. The construction industry is the biggest producer of MFRC trash because it uses bigger and thicker components, and this waste is generated both during manufacturing and at the end of their lives [5]. Since most research has concentrated on small-scale recycling, developing effective, scalable recycling technologies for these larger components is difficult [6].

MFRCs include basalt, wollastonite, mica, talc, and rockwool, among others, embedded in polymer matrices [7, 8]. Composites made from different types of fiber have unique qualities that make them useful in many different fields like automotive, aerospace, construction, and electronics, due to their ability to meet demanding performance requirements [9, 10]. Chemical stability, high tensile strength, and heat resistance make basalt fibers ideal for car dashboards and body panels [11, 12], Due to its dielectric and thermal insulation, mica and talc are used in circuit boards and insulating components, while wollastonite enhances fire-resistant construction materials [13]. In the past, asbestos was also used for comparable purposes; however, it is now mostly illegal to use because of the significant health hazards it poses, including as mesothelioma and lung cancer, its use has been heavily restricted or banned in many countries [14]. Filament type (continuous fibers, short fibers, particles, etc.) and polymer matrix type (epoxy, polyester, polypropylene, etc.) are two ways MFRCs are classified. For structural applications, continuous fibers provide the best combination of strength and stiffness, whereas for non-structural purposes, short fibers or particles offer better toughness and impact resistance. Potential for environmental sustainability is one of the main benefits of MFRCs. In comparison to synthetic fibers, basalt fibers have a lower production energy footprint, are completely recyclable, and are readily available. Roofing panels and insulation boards made of MFRCs are becoming more popular in eco-friendly building techniques. However, there are still obstacles to establishing comparable performance across applications, such as fiber property discrepancies and the necessity for standardized processing approaches [8]. Moreover, mineral fibers are more eco-friendly than synthetic ones, although there is still some damage to the environment from their extraction and processing. Reusing and recycling these materials have the ability to drastically cut down on trash and save valuable resources. The recovery of fibers and polymer matrices from end-of-life composites is made possible by techniques such as pyrolysis and solvolysis, which contribute to the shift towards a circular economy [15]. Figure 1 illustrates the estimated volume of FRPCs composite waste generated across various industries, with the construction sector being the largest contributor to FRPCs waste. For instance, a strong waste management plan that places an emphasis on recycling and resource recovery is necessary since construction waste is still the leading cause of MFRCs waste. The industry may achieve global sustainability objectives and fully use MFRCs as a future-proof, high-performance material by tackling these difficulties.

2 Classification of Natural Mineral Fibers

For polymer composites, mineral fibers are a necessary reinforcing material due to their high mechanical strength, resistance to heat, and environmental friendliness. Fibers like this may be found in nature or undergo a series of modifications to make them more useful in certain contexts. The manufacturing techniques and uses of basalt, mica, wollastonite, talc, and asbestos fibers in polymer composites are discussed in this section. In sectors that need high-performance materials, the various fiber types play a crucial role due to their distinct contributions. Recognized for their

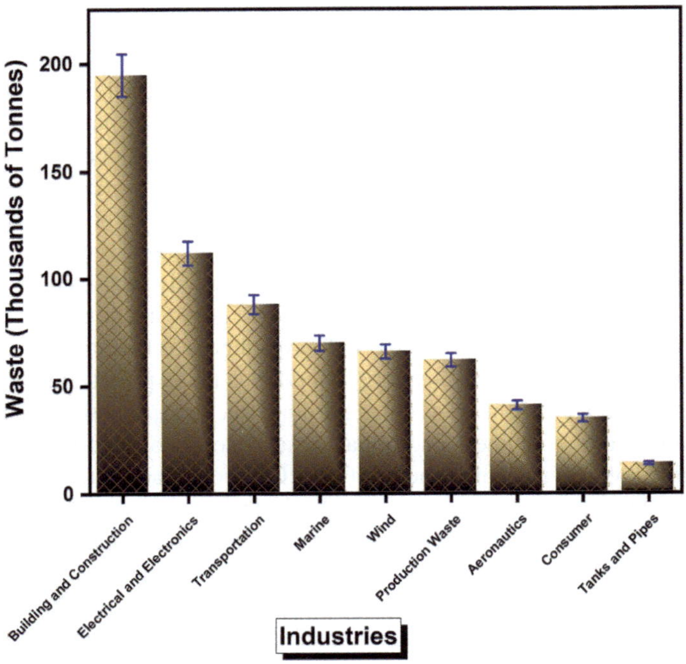

Fig. 1 Estimated generation of FRP waste (in thousands of ton) for the year 2025

adaptability and significance in contemporary material research, the several kinds of minerals used as reinforcing in polymer composites are shown in Fig. 2.

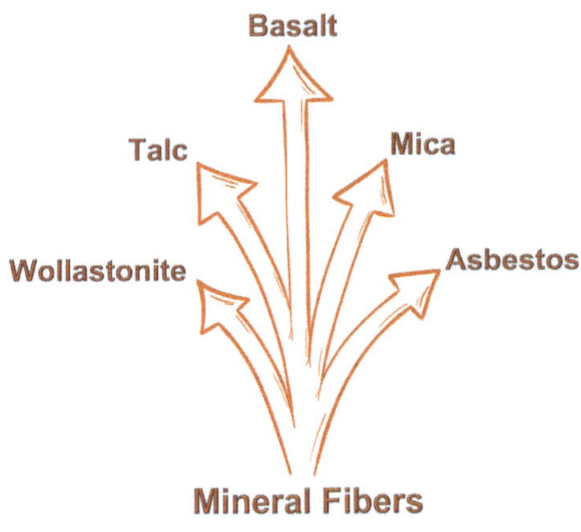

Fig. 2 Minerals used as reinforcing in polymer composites and their classification

2.1 *Basalt Fibers*

Basalt fibers are made from easily available and sustainably sourced basalt rock, which is formed during volcanic eruptions. The molten product is made by melting basalt rock at 1300–1400 °C. Depending on its function, the molten material is extruded into continuous fibers, which may be chopped or stranded.

2.2 *Manufacturing Process*

(a) Quarrying: Basalt rock is formed from deposits of volcanic activity.
(b) Melting: In a furnace, the basalt rock is heated to temperatures ranging from 1300 to 1400 °C.
(c) The process of fiber formation involves moving the molten basalt through spinnerets, which results in continuous fibers. It is common practice to freeze and wind these fibers onto spools before using them in composites.

Reinforced concrete, marine structures, and fire-resistant composites all benefit greatly from basalt fibers, which provide a number of advantages over glass and carbon fibers. Basalt fibers are quickly replacing synthetic ones, including carbon and glass, in industries that are trying to lessen their impact on the environment. Building, transportation, aviation, and wind power are just a few of the many industries that rely heavily on basalt fibers because of their exceptional suitability for environmentally conscious applications. Their excellent reputation for high tensile strength, resilience to heat, chemical stability, and environmental friendliness [16] is the foundation of their reliability in many high-performance applications.

2.3 *Mica*

Because of its exceptional dielectric properties and thermal insulation, the silicate substance mica is highly sought after. The naturally occurring mineral mica, when processed into sheets or fibers, offers a wide range of possible applications. Because of its inherent sheet-like structure, mica is an excellent insulator, thermal protector, and component in high-voltage electrical systems.

Manufacturing Process

(a) Mining for mica: The mica is extracted from pegmatite deposits and then carefully sorted along the planes of its natural cleavage.
(b) Fiber Formation: Mica is crushed into fine fibers or processed into sheets to make it ready for its many purposes.
(c) Processing: Mica fibers are often used in electrical and thermal applications, such as for weaving or compounding into polymer composites.

Transistors, capacitors, and circuit boards are only a few of the many devices that benefit from mica fibers' electrical insulating characteristics [17]. Their great dielectric strength and strong thermal stability make them ideal thermal protectors for usage in construction and aerospace materials.

2.4 *Wollastonite*

Wollastonite is a naturally occurring calcium silicate mineral that is immediately identifiable by its needle-like fiber structure. Reinforcing polymer composites using it increases their mechanical strength, flame resistance, and dimensional stability.

Manufacturing Process

(a) The extraction process begins with the grinding of the metamorphic rocks containing wollastonite into smaller pieces.
(b) Fiber Formation: After crushing the wollastonite, it is heated to temperatures about 1000 °C to organize the mineral's crystals into long, needle-like threads. This method increases the fiber's strength and reinforcing capabilities.

Many building products, including cement and fire-resistant boards, and automobile components, such dashboards, make use of wollastonite. Particularly useful in high-performance materials tailored to harsh settings, it improves composites' mechanical characteristics, heat resistance, and flame-retardant attributes [18, 19].

2.5 *Talc*

Talc, a mineral composed of magnesium silicate, is often used into polymer composites with the purpose of enhancing their rigidity, resistance to heat, and thermal conductivity. Composites made of thermoplastics and thermosets often have talc fibers added to them as reinforcements or fillers to boost their performance.

Manufacturing Process

(a) Extraction: Deposits of talc are extracted by mining.
(b) Grinding: After being finely crushed into a powder, raw talc may be used as a reinforcing particle or transformed into fibers.
(c) Modification: Particular processing methods may also improve the compatibility of talc with polymer matrices and optimize its fiber structure.

In addition to its use in building materials, packaging, and automobile components, talc fibers have many other practical applications. They contribute to the improvement of thermoplastics' mechanical qualities and processing features, leading to longer material life and better efficiency [20, 21].

2.6 Asbestos Fibers (Historical Use)

Fibrous silicate minerals like asbestos were highly prized for their strength, thermal stability, and tolerance to high temperatures in bygone eras. Many industrial products made use of it, including gaskets, brake pads, insulation, and roofing. Nevertheless, because to its significant health concerns, such as mesothelioma and lung cancer, its usage has been heavily limited or outright banned in some nations. Due to its heat and fire resistance, asbestos was widely utilized in cement sheets and brake linings and gaskets. However, breathing asbestos fibers is dangerous, thus its usage has been mostly abandoned [22, 23].

3 Environmental Implications of the MFRCs

The MFRC lifecycle, from raw material extraction and composite fabrication to energy use, recycling, and final disposal, contributes to diverse environmental impacts. High mechanical, thermal, and electrical qualities of MFRCs make them very desirable, yet their manufacturing and use are not without significant environmental costs. Important raw material extraction processes include basalt, mica, wollastonite, and talc, which create substantial environmental hazards. In example, the energy-intensive procedures used in mining for talc and mica may cause dangerous emissions, water pollution, deforestation, and land degradation. Reaching a balance between the advantages of MFRCs and sustainability objectives requires careful management of these aspects. While basalt is less damaging, its extraction and transportation still contribute to carbon emissions, and wollastonite processing also generates significant environmental costs [24, 25]. The recycling potential of MFRCs is limited, exacerbating their environmental impact. Basalt, mica, and wollastonite fibers are difficult to recycle through conventional methods, especially when embedded in complex composites. While some recycling technologies, like thermal and chemical methods, are under development, they are not yet commercially viable [26–28]. Additionally, even when fibers are separated from the polymer matrix, their recycled performance is often inferior to virgin fibers, limiting their reuse in high-performance applications [29].

However, the disposal of MFRCs at the end of life (EOL) presents significant environmental challenges shown in Table 1. While fibers such as basalt, mica, wollastonite, and talc are generally inert and non-toxic, the polymer matrix that binds these fibers can release harmful substances when subjected to degradation or incineration. Improper disposal of these composites can result in the accumulation of non-biodegradable materials in landfills, as polymer resins, often derived from petrochemicals, do not degrade naturally and may leach toxic by-products into the surrounding environment [33]. Furthermore, the long service life of mineral fiber-based products in sectors such as construction and automotive leads to delayed EOL disposal, contributing to the ongoing build-up of waste. Demolition and renovation activities, particularly in large-scale infrastructure projects, generate substantial waste, and the difficulty in recovering and recycling these materials exacerbates the environmental burden [34].

The environmental implications, including as carbon emissions and energy consumption, are exacerbated throughout the lifespan of MFRCs, encompassing manufacturing, transportation, and disposal. Basalt, mica, wollastonite, and talc fibers are more ecologically sustainable than synthetic alternatives such as glass or carbon fibers; yet, challenges persist regarding their extraction, processing, and end-of-life management. They are especially detrimental to the environment because to the energy necessary for their production, the absence of appropriate recycling options, and the complications associated with their disposal. Although these materials are very successful in the near term, the advancement of more efficient recycling techniques is essential for their sustained use in the future. Efficient recycling operations are essential to mitigate carbon emissions, redirect materials from landfills, and minimize waste. To mitigate the environmental impact of MFRCs while enhancing

Table 1 Comparative analysis of FRPCs waste disposal routes: strengths, limitations, and key considerations [30–32]

Disposal route	Strengths	Limitations	Key considerations
Landfilling	Cost-effective and straightforward implementation	Contributes to environmental pollution (dust, local air quality) and greenhouse gas emissions (methane) Long-term environmental impact due to toxic waste build-up	Limited available space in landfills, strict governmental regulations, and the risk of toxic waste accumulation impacting health and ecosystems
Incineration	Reduces waste volume, with potential for energy recovery	No material recovery; produces 50% residual ash waste Inefficient for recycling composite materials like GFRP (glass fiber-reinforced polymers)	Expensive to build, operate, and maintain; air pollutants resulting from combustion; inefficiency in recovering valuable materials like fibers
Co-incineration in cement kilns	Simultaneous material and energy recovery; uses FRP waste as fuel Efficient, scalable, and avoids ash generation	Loss of original fiber integrity; additional energy is required to reach high temperatures	Suitable for glass fiber-reinforced polymers (GFRP) only; the process necessitates size reduction of FRP waste; emissions of pollutants and particulates

their ecological benefits, it is essential to optimize their recycling processes for reintegration into the production cycle.

4 Recycling Strategies for MFRCs

Because of their intricacy and the strong interaction between mineral fibers and polymer matrices, recycling MFRCs is no easy task. If we want them to last longer and have less of an influence on the environment, we must ensure that we recycle properly. Mechanical, thermal (including pyrolysis), chemical, fluidized bed, and microwave-assisted recycling technologies have all been developed. There are advantages and disadvantages to each approach. There are primarily three types of recycling strategies for FRPCs, as shown in Fig. 3: mechanical, chemical, and thermal. Various thermal techniques, such as microwave treatments, fluidized bed recycling, and high-temperature pyrolysis, all play a part in the overarching objective of reducing waste while maximizing material recovery.

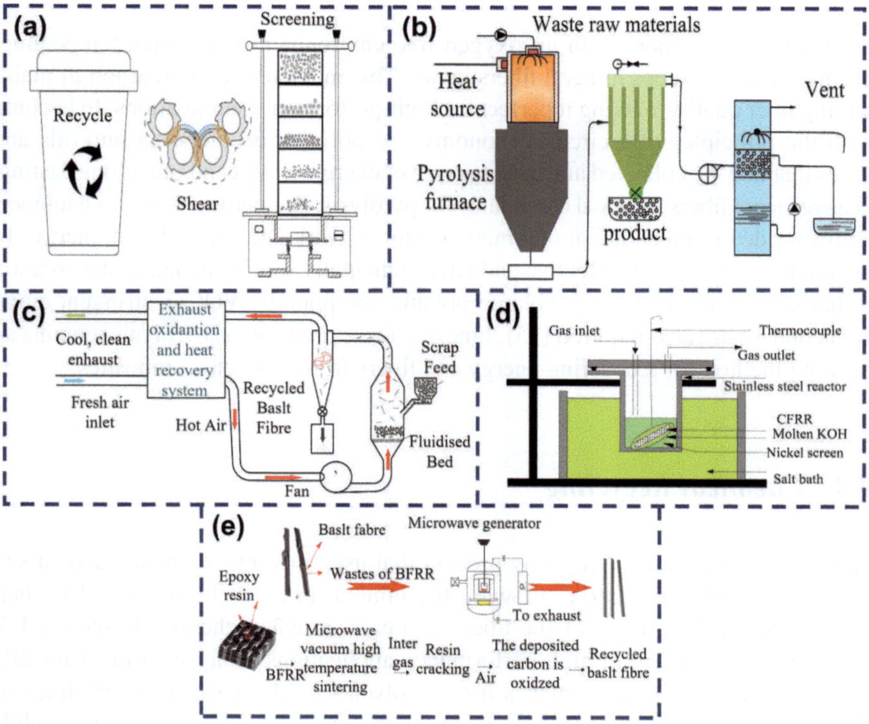

Fig. 3 Procedure diagrams for recycling trash from polymer composites **a** mechanical, **b** pyrolysis at high temperatures, **c** pyrolysis in a fluidized bed, **d** chemical breakdown, and **e** pyrolysis using microwaves

4.1 Mechanical Recycling

Many MFRCs are processed by mechanical recycling since it is the most common and easiest way. A physical process is used to physically break down composite materials into smaller particles, allowing the mineral fibers to be separated from the polymer matrix. Isolating the fibers usually involves shredding or grinding the composite into smaller bits, followed by techniques like air classification or vibration sifting. This technique has certain limits, but it is cost-effective and ideal for processing big amounts of garbage. Degradation of the fibers occurs often during the process, leading to decreased strength and improper use for high-performance applications. Also, the fibers' quality might be compromised if leftover polymer matrix material remains connected to them. Notwithstanding these obstacles, mechanical recycling works well for insulation and other non-structural building materials when the recovered fibers' mechanical qualities are less important [28].

4.2 High Temperature Pyrolysis

Decomposing composites in an oxygen-free environment using high-temperature pyrolysis, which keeps mineral fibers intact. This method does a great job of maintaining fiber quality, making it perfect for high-performance applications. In keeping with the principles of a circular economy, the polymer decomposes into oils and gases that may be collected and used to recover energy. Basalt and other long-lasting, heat-resistant fibers are ideal candidates for pyrolysis. But getting to the high temperatures needed to break down polymers is a lot of work and uses a lot of energy. To guarantee the process is effective and environmentally safe by avoiding the creation of harmful by-products such volatile organic compounds (VOCs), stringent emission controls are also required [35]. Ignoring these drawbacks, pyrolysis remains an exciting method for recovering energy and fibers from obsolete composites.

4.3 Chemical Recycling

Solvolysis is a chemical recycling process that uses solvents or chemical agents to dissolve the polymer matrix, allowing the mineral fibers to be recovered in their entirety. Since the quality of the fibers is preserved when chemically recycled, it is better to reuse them in high-performance applications rather than mechanically recycle them. This process makes use of solvents, such as supercritical fluids or alkaline solutions, to dissolve the polymer matrix in a selective manner, all the while keeping the fibers intact. By effectively isolating and recovering high-quality fibers, this technique offers a more controlled and precise method of recycling. Chemical recycling does have some challenges, however, including the high cost of solvents, the inconvenience of solvent recovery systems, and the difficulty of scaling up the process to meet the needs of large businesses. The procedure becomes much more intricate and costly due to the usage of solvents, which may cause injury if not handled correctly. Notwithstanding these challenges, chemical recycling has great promise for producing recovered materials of good quality for demanding applications [36].

4.4 Fluidized Bed Recycling

Fluidized bed recycling is an innovative technique that uses hot gases or air to break-down the polymer matrix of composite materials. A fluidized bed reactor is used for this purpose. Since it does not operate at the high temperatures required by conventional pyrolysis, the process is gentler on the fibers and uses less energy while yet preserving their mechanical properties. This process involves feeding composite materials into a reactor, which uses hot gas or air to break down the polymer matrix while the mineral fibers stay intact. We then separate the fibers for future usage after air-collecting them. The energy efficiency and environmental friendliness of this system are superior than those of more conventional thermal recycling methods. But it could be expensive and resource-intensive to scale it up for use in industry. In addition, different types of composites employ different combinations of polymer and mineral fibers, therefore optimizing the method for these types of composites will need more research and development [37].

4.5 Microwave Assisted Recycling

A novel approach to recycling, microwave-assisted recycling breaks down fiber-reinforced composites by selectively heating and decomposing the polymer matrix. The technique focuses on the polymer matrix rather than the mineral fibers because of how well the matrix absorbs microwave radiation. The polymer breaks down, but the fibers remain intact. Among the many advantages of this technology is its ability to speed up recycling cycles and reduce energy consumption (since the matrix is heated directly by the microwaves, rather than the fibers). Plus, it slows down fiber degradation, so their mechanical properties are preserved for use again in high-performance settings. Nevertheless, while microwave-assisted recycling shows promise, it is still in the experimental stage and has to overcome challenges like scalability and precise process control before it can be used in industrial settings [38].

4.6 Hybrid Method

The shortcomings of individual recycling procedures are being addressed by the rise of hybrid recycling technologies. These approaches integrate two or more techniques. These technologies improve recycling efficiency, increase fiber recovery rates, and retain the quality of recovered fibers by merging mechanical, thermal, and chemical approaches. For instance, chemical recycling in conjunction with mechanical procedures may provide greater fiber separation without sacrificing integrity, and pyrolysis combined with mechanical separation can generate high-quality fibers by degrading the matrix. Innovative approaches have shown remarkable fiber recovery with low tensile strength loss. One such approach is the catalyst-free method developed by Zabihi et al., which utilizes tartaric acid and hydrogen peroxide. Similarly, glass fibers recovered from epoxy-based composites using a microwave treatment and solvolysis hybrid have qualities similar to virgin fibers [39]. Despite their promise, hybrid approaches need the integration of several technologies, making them more complicated and expensive than single-process procedures. Addressing these problems, streamlining processes, and reducing costs is crucial for making hybrid recycling technologies practicable for large-scale applications, which in turn requires ongoing research and development.

Using important performance indicators including processing temperatures, energy consumption, and global warming potential (GWP), Table 2 compares and contrasts different recycling techniques for fiber-reinforced composites (FRCs). It also shows the pros and cons of each process, illuminating important details like fiber recovery efficiency, resin separation efficacy, and environmental impact. Insights into the sustainability, cost-effectiveness, and feasibility of various recycling solutions are provided by comparing these features; this, in turn, helps to determine the most viable ways for handling FRC trash.

4.7 Novel Recycling Techniques

The difficulties of recycling FRPCs have recently been addressed by the development of new recycling processes, which provide energy-efficient, selective, and environmentally benign alternatives to the conventional approaches (see Fig. 4). For carbon fiber-reinforced polymer composites (CFRPCs) and polyester-based composites, superheated steam recycling, which operates at 500 °C to 700 °C, efficiently breaks down polymer matrices, allowing for the recovery of high-quality fibers [41]. The use of electric currents in electrolyte solutions, such as NaCl, has allowed for the electrochemical recycling of CFRPs to achieve fiber recovery of up to 80% of virgin fiber tensile strength and increased recovery for glass fiber-reinforced polymer composites (GFRPCs) [42]. Fiber recovery efficiency of polyester-based GFRPs is improved compared to mechanical recycling using this method, and CFRP fibers

Table 2 Evaluation of recycling techniques for FRPCs [40]

Method	Recycled fiber performance	Processing temperature (°C)	Energy consumption (MJ/kg)	Global warming potential (CO_2 eq/kg)	Resin separation rate (%)	Advantages	Disadvantages
Mechanical recycling	40–50%	Room temperature	1.10–1.5	0.10–1.5	70	Low cost, simple process, minimal pollution	Low fiber recovery rate, fiber damage
High-temperature pyrolysis	25–50%	350–800	30–40	2.5–3.0	67	High processing capacity, wide applicability	High energy use, emissions from waste gases
Fluidized bed pyrolysis	50%	400–600	9.0–11	1.5–1.8	70	Clean fiber recovery, efficient process	Fiber degradation, energy-intensive process
Microwave pyrolysis	70–80%	300–600	8.0–10	1.3–1.6	100	Fast processing, excellent resin separation	High setup and operational cost
Chemical recycling	85–95%	200–450	35–40	1.4–1.6	98–99	High fiber recovery rate, resin recovery possible	Complex process, environmental impact

with up to 80% of virgin fibers' tensile strength may be recovered [43]. Using high-frequency sound waves, ultrasonic recycling speeds up the breakdown of resins, allowing for the recovery of fibers. Thermoplastic composites have the potential to be pre-cracked using ultrasonic waves in a way that is safe for the fibers. A method that involves ultrasonic treatment in conjunction with chemicals like nitric acid and hydrogen peroxide has been shown to effectively restore CFRPC fibers while minimizing strength loss [44]. Ionic liquid recycling, which comprises mildly dissolving polymer matrices in ecologically friendly solvents, may restore more than 95% of the original tensile strength of the fibers used in epoxy-based CFRPCs [45]. A recent advancement in this area is the use of sulfuric acid to dissolve the epoxy matrix in GFRPCs, resulting in valuable industrial by-products [46]. Finally, biological recycling is a very young but promising field of study that employs enzymes or bacteria to degrade polymer matrices. Enzymes have recently emerged as a viable tool in the fight against polyurethane resins and their valuable by-products [47]. The Bizente project is an EU initiative that is one of many investigating engineered enzymes that may biodegrade thermoset composites; this could lead to a more environmentally friendly method of recycling FRPCs [48].

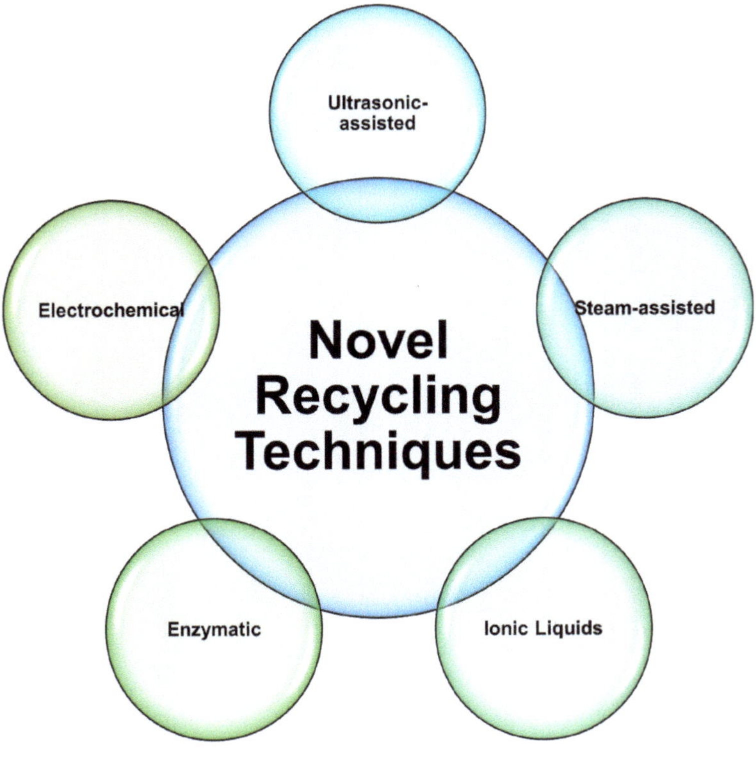

Fig. 4 Energy drivers, activators, and materials in innovative recycling approaches

5 Studies on the Recycling and Reuse of Recycled Fibers in Polymer Composites

Recent studies have explored various methods for recycling basalt fiber-reinforced composites, focusing on preserving fiber integrity and mechanical properties. One such method is solvolysis, a chemical recycling process that dissolves the polymer matrix to recover the embedded fibers. Persico et al. demonstrated the effectiveness of solvolysis in recycling basalt fiber-epoxy laminates at temperatures below 160 °C, successfully separating the fibers without significant degradation [49]. Figure 5 showed the process scheme for recycling method adopted.

The study investigates the recycling of basalt fiber-reinforced composites through solvolysis, focusing on the mechanical properties and surface characteristics of the recycled fibers. Visual analysis confirmed that the recycled woven basalt fabrics were resin-free, exhibiting a similar appearance to virgin fibers. Mechanical testing showed that recycled fibers retained 90.5% of the original breaking tenacity compared to virgin fibers, though some reduction in mechanical properties was observed. This decline may be due to surface defects from handling or partial removal of fiber sizing during solvolysis, which plays a critical role in improving fiber-matrix bonding and reducing friction. SEM analysis revealed that the surface quality of recycled fibers was comparable to virgin fibers, with smooth lateral surfaces and minimal residual resin. However, surface flaws and fracture patterns indicative of brittle failure were observed, consistent with the characteristics of ceramic fibers. Fracture nucleation sites were linked to both surface and internal impurities, highlighting the impact of the recycling process on fiber integrity, and all results were shown in Fig. 6. Overall, the study demonstrates that solvolysis offers a promising method for recycling basalt

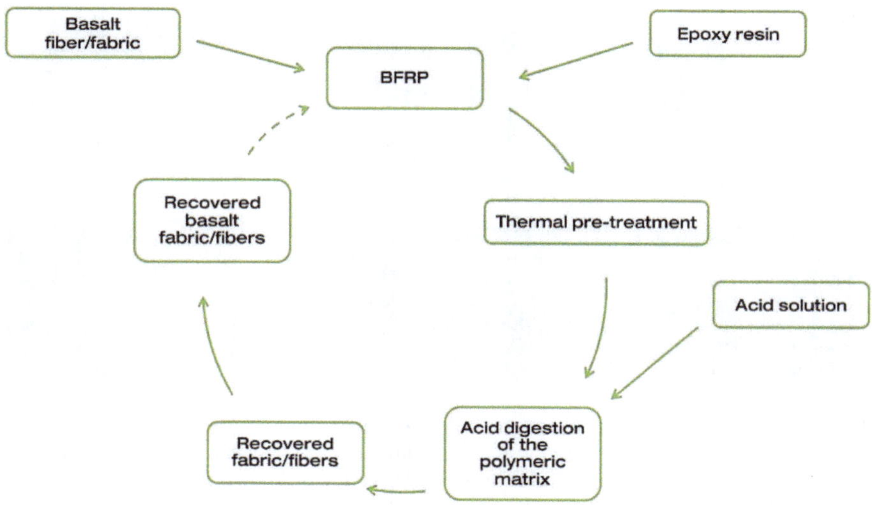

Fig. 5 Recycling of basalt fibers through the solvolysis process [49]

fibers with minimal loss in mechanical properties, though further research on surface regeneration is needed [49].

Thermal recycling techniques, such as pyrolysis, have also been investigated. Bhat et al. examined the properties of thermally recycled basalt fibers and their composites, finding that while some mechanical properties were retained, there was a noticeable reduction in tensile strength compared to virgin fibers [50]. The study investigates the impact of heat treatment on the tensile properties of basalt fibers, focusing on the effects of temperature and heating time on failure stress. The results show that while heat treatment does not significantly alter the Young's modulus or stiffness of the fibers, the failure stress decreases with increased temperature and time. Basalt fibers treated at temperatures above 350 °C experience significant loss in strength due to the degradation of the organic sizing agent, which reduces fiber-matrix bonding. XRD and TGA revealed no crystallization or significant mass loss up to 600 °C, with the primary loss in strength attributed to surface flaw growth. SEM analysis confirmed that failure always initiated at the surface, with larger flaws developing at higher temperatures. When recycled basalt fibers were used in polymer composites, their stiffness remained largely unaffected, but tensile strength decreased due to the loss of the sizing agent. Composites with recycled fibers showed similar strength loss as

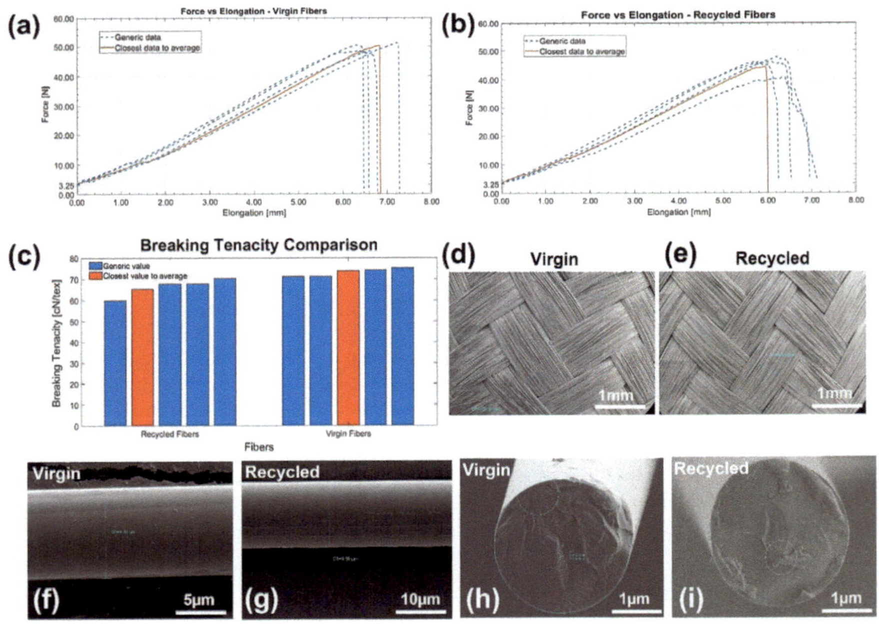

Fig. 6 **a, b** Force vs. elongation curve compares the mechanical properties of recycled and virgin basalt fibers, **c** while breaking tenacity shows the differences in strength. **d, e** The woven fabric of recycled fibers is resin-free and similar to virgin fibers. **f, g** SEM images of the lateral surfaces reveal smooth fiber surfaces, with recycled fibers showing minor surface flaws. **h, i** Fracture analysis indicates nucleation sites on both the surface and within the bulk of recycled fibers, highlighting the impact of recycling on fiber integrity [49]

the single fibers, particularly above 350 °C. Overall, while thermally recycled basalt fibers maintain their stiffness, their tensile strength is significantly reduced due to thermal degradation and surface flaw growth [50].

Mechanical recycling methods involve grinding the composite material to reclaim fibers, but this often results in reduced fiber length and compromised mechanical properties. Finnerty et al. evaluated the impact of mechanical recycling on polylactic acid (PLA) and PLA-based natural fiber-reinforced composites, including those with basalt fibers. The findings indicated that multiple recycling cycles led to a decline in impact strength and tensile properties [51]. Moreover, this study revealed that the mechanical properties of PLA-based natural fiber-reinforced polymer composites (NFRPCs) degrade with increasing recycling steps. After three recycling cycles, the impact strength decreased by 20–40%, and tensile strength dropped by 40–50% for basalt fiber (BF)-based composites, while HNT-based composites showed a smaller reduction of 15–25%. Discoloration, indicated by a ΔE^*ab value greater than 5, was observed due to thermal degradation. SEM analysis confirmed a significant reduction in fiber length, especially in BF composites, where fiber length decreased by up to 50% after six recycling steps. Despite these declines, HNT-based composites exhibited better retention of mechanical properties shown in Fig. 7, suggesting the importance of fiber type and surface adhesion in recycled composites.

Innovative approaches are being developed to enhance the recycling and reuse of basalt fiber-reinforced composites. For instance, integrating artificial intelligence into the recycling process has been proposed to improve efficiency and material recovery rates. A recent study discussed the potential of AI technology in optimizing the recycling of basalt fiber-reinforced resin materials, aiming to address environmental concerns and resource utilization [52]. These studies collectively contribute to the advancement of sustainable practices in the lifecycle management of basalt fiber-reinforced composites, highlighting the importance of developing effective recycling methods to maintain material performance and support environmental sustainability.

However, natural minerals like mica, talc, wollastonite, and rockwool have peculiar characteristics that make recycling their fibers from polymer composites a formidable task. Because of their fragility and difficulty in separating from the polymer matrix, mica fibers, for instance, often degrade and have their length reduced during mechanical or thermal recycling. Similarly, talc fibers have problems with adhesion between the matrix and the fibers and also shorten when recycled, which reduces their reinforcing power. Thermostatically recycled wollastonite fibers are more robust and durable, although they still have issues with quality retention and viable fiber separation. Rockwool fibers are notoriously difficult to recycle because of their brittleness, which makes recovery techniques complicated, energy intensive, and results in massive fiber loss. To overcome these drawbacks fiber brittleness, low separation efficiency, and material property loss and increase the recyclability and reusability of these mineral fibers in composite applications, new surface treatment technologies and state-of-the-art recycling methods are required.

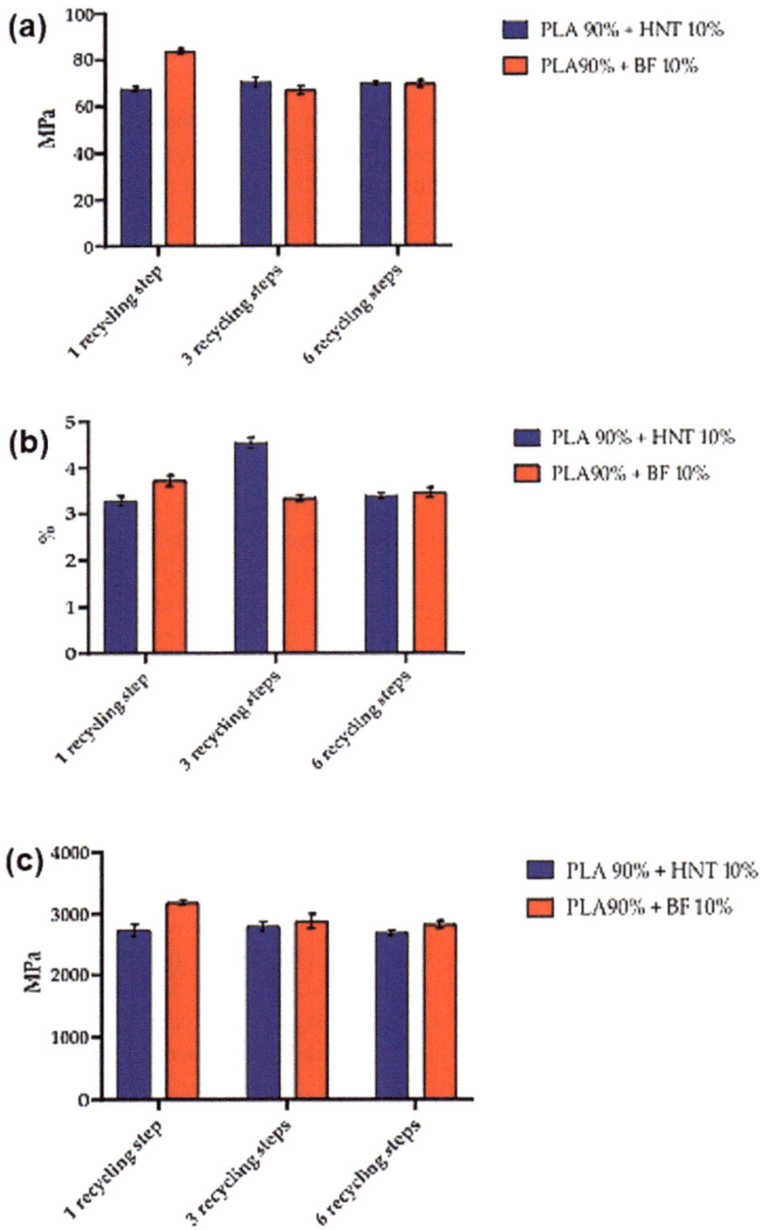

Fig. 7 Tensile characteristics of all six samples that were tested: The three variables in question are **a** stress at break, **b** strain at break, and **c** Young's modulus. There were two samples that were tested: one with 90 weight percent PLA and 10 weight percent HNT, and the other with 90 weight percent PLA and 10 weight percent BF. Bars representing errors show standard error of the mean [51]

6 Evaluation of the Life Cycle Impact of Composite Recycling

Recycling composites and the accompanying costs have been the subject of extensive research using LCC analysis, which considers all expenditures incurred by a product from its inception to its eventual disposal. Thoroughly analyzing the expenses of the process and comparing various disposal techniques is essential for recycling composite materials. Recyclability and product lifecycle management for composites are interdependent, as seen in Fig. 5. Virgin fiber composites are a major source of energy consumption, air pollution, and waste management problems due to the materials and energy needed for their creation, as well as their manufacturing and disposal procedures. In light of these effects, it is clear that the composite lifespan needs sustainable and cost-effective recycling solutions. The LCA framework has been developed to evaluate the energy consumption and environmental footprint of a product or process. This methodology is validated and standardized by the international organization for standardization (ISO) in the ISO 14040 and 14,044 standards [53]. LCA evaluates the entire life cycle of a product or process, starting from raw materials and ending with disposal. Numerous studies using LCA have been conducted to assess the recycling of composites in terms of energy and environmental impacts, predicting a promising future for their life-cycle during both application and recycling phases [54, 55]. Figure 8 identifies the potential end-of-life product routes for composite materials.

- Reuse: Reusing composite materials may not be feasible for some applications. For this route, manufacturing processes must be adapted to facilitate material reuse.

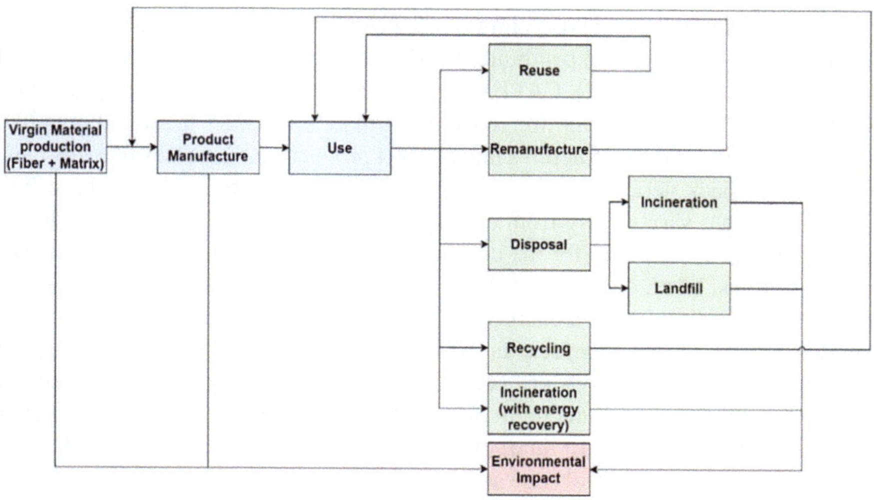

Fig. 8 Life cycle assessment and recycling pathways of FRPCs

- Remanufacture: Some composite products can go through restoration processes to regain their original properties and performance.
- Disposal: This process refers to eliminating waste without recovering any value.
- Recycling: Recycling involves recovering value from end-of-life products at a material level. Unlike remanufacturing or reuse, recycling involves reproducing the form of the product and reprocessing materials. Incineration with energy recovery is not considered a part of recycling since it does not result in the recovery of material.

These pathways for fiber-reinforced composites are crucial, as different studies reviewed in this article focus on various recycling or disposal routes, including the supply chain [56]. For example, Farel et al. developed a model to assess the feasibility of glazing recycling [57], while Witik et al. focused on LCA for energy recovery through incineration [58]. The literature reviewed reveals that the waste treatment stages in the lifecycle of fiber-reinforced composites have been inadequately studied. Most works that adopt LCA emphasize the production phase more than utilization [58, 59], but previous studies have also focused on different stages of composite products. For instance, Meng et al. proposed a cost model that considered three phases: (a) the fluidized bed recycling phase, (b) the processing and manufacturing of recycled carbon fiber (rCF) automotive components, and (c) the application phase, which considers weight savings in automotive components [60]. In contrast, Li et al. compared the financial viability of mechanical recycling against landfilling and incineration, concentrating only on the recycling phase [61]. The study specifically focused on assessing recycling and disposal routes without including costs incurred during the production and application phases. The scope is limited to studies that examined cost models for the recycling phase, which may also include cost information for other phases. Different cost categories associated with the recycling of composite materials have been clarified for better understanding.

The costs incurred during recycling can be categorized as: (1) capital expenditure (CAPEX) and operational expenditure (OPEX), (2) variable and fixed costs, and (3) direct and indirect costs [62]. CAPEX are the initial investments needed to acquire fixed assets, which are crucial for starting a firm. These assets include things like machinery, tools, and the structures that support production. Expenses like this have a lengthy payback period and eventually disappear [63]. The CAPEX in recycling operations might vary greatly from one technology or approach to another. On the other side, OPEX accounts for the continuous expenditures of running a business, such as utilities, labor, operating costs, and overhead. Depreciation, quality control, insurance, and administrative charges are indirect costs that add up to overall operating costs [62]. Direct expenses, such personnel, materials, and maintenance, are associated with distinct goods or operations and are therefore easier to allocate, in contrast to indirect costs, which are not directly attributable to anything in particular. It is common practice to think of direct costs as being variable and subject to change when production levels rise and fall, and indirect costs as being constant and unchanging as output falls. It is crucial to distinguish between these expenses when evaluating the financial viability of recycling operations [64].

7 Current Trends in Cost Modeling for Recycling FRPCs

There have been a lot of studies looking at the financial viability of recycling FRCs using LCC models. These models assess the overall costs of a product throughout its entire lifecycle, from design to disposal, and they compare different recycling methods by taking both financial and environmental factors into account. One study by Li et al. looked into mechanical recycling of CF composites and discovered that it had high disassembly costs (1.38 GBP per kg) and a poor fiber recovery rate of just 40% [65]. They didn't include in the possibility of income from energy recovery during incineration, and they didn't include production and consumption phase expenses either, as these are constants regardless of the disposal mode. Compared to landfilling and incineration, which have lower costs, the results indicated that recycling takes more money. In contrast, Meng et al. outlined four steps in their cost model for carbon fiber recovery using a fluidized bed process: CF recycling, processing rCF, manufacturing components from rCF, and application. The application phase encompassed potential fuel savings with lighter rCF components in automotive applications [60]. The cost of recovering CF was determined to be $5/kg by the researchers, which is just 15% of the price of virgin CF. This resulted in substantial price savings compared to commodities such as steel and aluminum. Other aspects that impact the process's overall economic viability, including as the costs of sorting, disassembly, and transportation of CF materials, may have been neglected in their analysis. Chemical, mechanical, and pyrolytic recycling methods were evaluated by Vo Dong et al. alongside supercritical water (SCW). They arrived at their findings by using significant economic criteria, including unit cost per mass of waste (UCW) and cost per recovered fiber (UCF) [54]. Mechanical recycling and pyrolysis are more cost-effective for CF waste than chemical processes like SCW, according to their research. According to the study, co-incineration in cement kilns is a method for recovering materials and energy, but it isn't as economical as landfilling or incineration. In their suggested closed-loop model, Hagnell and Åkermo highlighted the possibility of using recycled carbon fibers and glass fibers to save costs by as much as half. This might be particularly advantageous when substituting inexpensive materials like as virgin CF for structural purposes. The study emphasized the importance of fiber recovery and mechanical performance, demonstrating how recycling can offer substantial savings if higher recovery rates and quality fiber applications are achieved [65]. Implementing a cost–benefit analysis (CBA) to examine variables such recycled price changes, landfilling charges, and logistics, Farel et al. developed a model to evaluate the viability of recycling glazing materials from ELVs [57]. Relevant to the recycling of non-metallic materials, such as fiber-reinforced composites, their results provide useful clarifications. Just like that, Zhao et al. presented a cost–benefit analysis of disassembling and recycling airplane components as an economic indicator. Their research showed that aircraft engines were more economically viable to recycle owing to their high salvage value; however, recycled aircraft bodies were less so owing to the high expenses of dismantling and processing [66]. These studies highlight the future potential for improved economic feasibility due to technological advancements, higher recovery rates, and a higher market value for recycled fibers, despite the fact that composite recycling still faces financial challenges, especially with regard to transportation and pre-treatment costs.

8 Future Directions

MFRCs recycling offers a great chance to cut down on waste and save resources, but it also has a lot of economic and environmental problems. In order to maximize fiber recovery while minimizing energy consumption, future efforts should focus on developing efficient and scalable recycling technologies. This should include refining procedures like fluidized bed recycling, pyrolysis, and microwave-assisted approaches. It is believed that mechanical, thermal, and chemically integrated hybrid recycling systems may significantly increase efficiency. Recycling fibers while maintaining their mechanical characteristics is crucial, especially for degradable materials like talc and mica. To improve the quality of fiber for high-performance uses, research into surface treatments and regeneration methods can be crucial. There is an urgent need for further research on cost-effective recycling models that take into account cutting-edge technology like AI-assisted sorting and cover every step of the process, from extracting raw materials to disposing of them. The economic and environmental effects of recycling programs can only be determined by comprehensive life cycle assessments, such as LCC and LCA. In order to maximize recovery rates and efficiency, it is necessary to create fiber-specific recycling processes. This includes basalt and mica. It would be wise to delve more into emerging methods such as electrochemical and ultrasonic recycling, since these show great potential. To create uniform rules and execute efficient recycling solutions, it is essential that businesses, universities, and lawmakers work together. Improving the sustainability, cost-effectiveness, and integration of MFRC recycling into a circular economy may be achieved by focusing on these areas.

9 Conclusion

The increasing demand for FRPCs in sectors including electronics, construction, aerospace, and automotive has spurred the need to reduce their environmental effect. When compared to other composites, MFRCs have superior mechanical strength, thermal stability, and longevity. These cutting-edge materials are advantageous for high-performance uses due to their adaptability and composition, which includes mineral fibers such as basalt, mica, wollastonite, and talc, in either their native or transformed forms. The challenging problem of successful recycling lies in the strong connections that exist between mineral fibers and polymer matrices. mechanical, thermal (e.g., pyrolysis), chemical, and innovative approaches including fluidized bed, hybrid, and microwave-assisted recycling have all been investigated in an effort to overcome these challenges. Although mechanical recycling is easy and inexpensive, it often damages the fibers, making them unfit for uses requiring high performance. Although thermal processes, like as pyrolysis, may increase fiber recovery rates, they are highly energy demanding and need complex emission control systems. Though solvolysis and other chemical recycling methods show promise in maintaining fiber integrity, problems with solvent control and scalability limit their use. Though still in its experimental stages, new methods like as fluidized bed recycling and microwave-assisted recycling are promising for commercial usage due to their potential to reduce energy consumption without sacrificing fiber quality. Sustainability and cost-effectiveness of various recycling systems have been evaluated using LCA and LCC frameworks, which take into account both environmental and economic factors. The high operating expenses of recycling, especially for energy consumption, transportation, and pre-treatment operations, restrict its financial sustainability, despite the environmental advantages it may give. Considerations including the market demand for recycled fibers, the recycling technology used, and the kind of reinforcing material (basalt) all play a role in determining the economic feasibility. Some mineral fibers, such as talc and mica, are particularly brittle, which makes it difficult to separate the fibers from the polymer matrix and recover them. Another issue is that some fibers degrade during recycling, which lowers their chances of being used for high-performance purposes again. Reducing waste, conserving resources, and lowering the environmental impact of MFRCs may be achieved by recycling, especially when integrated into a circular economy framework. This is in addition to the benefits that recycling already provides.

References

1. Fiber Reinforced Polymer (FRP) Composites Market - Forecasts from 2023 to 2028 [Online]. Available: https://www.researchandmarkets.com/reports/5547603/fiber-reinforced-polymer-frp-composites?utm_source=GNOM&utm_medium=PressRelease&utm_code=q88slz&utm_campaign=1679794+-+Worldwide+Fiber+Reinforced+Polymer+Composites+Industry+to+2026+-+by+Type%2C+En
2. Markets Ra (2026) Fiber reinforced polymer (FRP) composites market—forecasts from 2021 to 2026. Accessed: 3–12–2024. [Online]. https://www.researchandmarkets.com/reports/5547603/fiber-reinforced-polymer-frp-composites?utm_source=GNOM&utm_medium=PressRelease&utm_code=q88slz&utm_campaign=1679794+-+Worldwide+Fiber+Reinforced+Polymer+Composites+Industry+to+2026+-+by+Type%2C+En
3. EC Observer Overview of the Global Composites Market, ed, 2022
4. Commission E (2018) Directive (EU) 2018/850 of the European Parliament and of the Council of 30 May 2018 Amending
5. Directive 1999/31/EC on the Landfill of Waste (2018) Off J Eur Union, vol L 150:100–108
6. Naqvi S, Prabhakara HM, Bramer E, Dierkes W, Akkerman R, Brem G (2018) A critical review on recycling of end-of-life carbon fibre/glass fibre reinforced composites waste using pyrolysis towards a circular economy. Resour Conserv Recycl 136:118–129
7. Gonçalves R, Martinho A, Oliveira JP (2022) Recycling of reinforced glass fibers waste: current status. Materials 15(4):1596
8. Fiore V, Scalici T, Di Bella G, Valenza A (2015) A review on basalt fibre and its composites. Compos B Eng 74:74–94

9. Lei Y et al (2024) Natural flame retardant minerals for advanced epoxy composites. Fire 7(9):308
10. Mohamad Moasas A, et al (2022) Bibliographic trends in mineral fiber-reinforced concrete: a scientometric analysis. Front Mater 9:1100276
11. Bledzki A, Gassan J (1997) Natural fiber reinforced plastics ed: Marcel Dekker, Inc., New York, pp 787–809
12. Jamshaid H, Mishra R (2016) A green material from rock: basalt fiber–a review. J TextE Inst 107(7):923–937
13. Pareek K, Saha P (2019) Basalt fiber and its composites: an overview. In: Proceedings of national conference on advances in structural technologies (CoAST-2019), vol 1, no 1: Department of Civil Engineering National Institute of Technology Silchar, pp 53–62
14. Bazan P (2024) Hybrid synthetic/mineral fibers: composites and automotive applications. In: Synthetic and mineral fibers, their composites and applications: Elsevier, pp 513–527
15. Mir WAY, et al (2023) Asbestos-related diseases and its impact on health: an updated review article. Curr Pulmonol Rep 12(4):244–255. https://doi.org/10.1007/s13665-023-00324-x
16. Yang Y, Boom R, Irion B, Van Heerden D-J, Kuiper P, De Wit H (2012) Recycling of composite materials. Chem Eng Process 51:53–68
17. Jagadeesh P, Rangappa SM, Siengchin S (2024) Basalt fibers: an environmentally acceptable and sustainable green material for polymer composites. Constr Build Mater 436:136834
18. Shen J, et al (2023) An all-natural, bioinspired, biodegradable electrical insulating composite based on lignocellulose and mica tailings. Int J Biol Macromol 253:127222. https://doi.org/10.1016/j.ijbiomac.2023.127222
19. Chan JX, Wong JF, Hassan A, Mohamad Z, Othman N (2020) Mechanical properties of wollastonite reinforced thermoplastic composites: a review. Polym Compos 41(2):395–429
20. Nair NA, Sairam V (2021) Research initiatives on the influence of wollastonite in cement-based construction material-A review. J Clean Prod 283:124665
21. 얀준 (2015) Soundproofing and Mechanical Properties of Nanofiller-reinforced Polypropylene Composites," 서울대학교 대학원
22. Udebuani O, Onitiri M, Adewumi O (2021) Experimental and numerical determination of the heat transfer coefficient of fiber-glass/talc-filled epoxy
23. Virta RL (2002) Asbestos: geology, mineralogy, mining, and uses. Citeseer
24. Spasiano D, Pirozzi F (2017) Treatments of asbestos containing wastes. J Environ Manage 204:82–91
25. Jagadeesh P, Puttegowda M, Mavinkere Rangappa S, Siengchin S (2021) A review on extraction, chemical treatment, characterization of natural fibers and its composites for potential applications. Polym Compos 42(12):6239–6264.
26. Kumar KP, Sekaran ASJ (2014) Some natural fibers used in polymer composites and their extraction processes: a review. J Reinf Plast Compos 33(20):1879–1892
27. Asmatulu E, Twomey J, Overcash M (2014) Recycling of fiber-reinforced composites and direct structural composite recycling concept. J Compos Mater 48(5):593–608
28. Wu Z, Wang X, Liu J, Chen X (2020) Mineral fibres: basalt. In: Handbook of Natural Fibres: Elsevier, pp 433–502
29. Bernatas R, Dagréou S, Despax-Ferreres A, Barasinski A (2021) Recycling of fiber reinforced composites with a focus on thermoplastic composites. Clean Eng Technol 5:100272
30. Krauklis AE, Karl CW, Gagani AI, Jørgensen JK (2021) Composite material recycling technology—state-of-the-art and sustainable development for the 2020s. J Compos Sci 5(1):28
31. Gharde S, Kandasubramanian B (2019) Mechanothermal and chemical recycling methodologies for the Fibre Reinforced Plastic (FRP). Environ Technol & Innov 14:100311. https://doi.org/10.1016/j.eti.2019.01.005
32. Gonçalves RM, Martinho A, Oliveira JP (2022) Recycling of reinforced glass fibers waste: current status. Materials 15(4):1596 [Online]. https://www.mdpi.com/1996-1944/15/4/1596
33. Qureshi J (2022) A review of recycling methods for fibre reinforced polymer composites. Sustainability 14(24):16855

34. Azwa Z, Yousif B, Manalo A, Karunasena W (2013) A review on the degradability of polymeric composites based on natural fibres. Mater Des 47:424–442
35. Ray BC, Rathore D (2015) Environmental damage and degradation of FRP composites: a review report. Polym Compos 36(3):410–423
36. Abdallah R et al (2021) RETRACTED: a critical review on recycling composite waste using pyrolysis for sustainable development. Energies 14(18):5748
37. Chen R, Cheng L, Gu J, Yuan H, Chen Y (2023) Research progress in chemical recovery technology of fiber-reinforced polymer composites. CIESC J 74(3):981
38. Ateeq M (2023) A review on recycling technique and remanufacturing of the carbon fiber from the carbon fiber polymer composite: Processing, challenges, and state-of-arts. Compos Part C: Open Access 100412
39. Karuppannan Gopalraj S, Kärki T (2020) A review on the recycling of waste carbon fibre/glass fibre-reinforced composites: fibre recovery, properties and life-cycle analysis. SN Appl Sci 2(3):433
40. Zabihi O, Ahmadi M, Liu C, Mahmoodi R, Li Q, Naebe M (2020) Development of a low cost and green microwave assisted approach towards the circular carbon fibre composites. Compos Part B: Eng 184:107750. https://doi.org/10.1016/j.compositesb.2020.107750
41. Cai B (2024) Recycling and reuse of basalt fiber reinforced resin materials with artificial intelligence technology. J Electr Syst 20:2063–2074. https://doi.org/10.52783/jes.1656
42. Wada M, Kawai K, Suzuki T, Hira H, Kitaoka S (2016) Effect of superheated steam treatment of carbon fiber on interfacial adhesion to epoxy resin. Compos Part A: Appl Sci Manuf 85:156–162. https://doi.org/10.1016/j.compositesa.2016.03.024
43. Oshima K, Matsuda S, Hosaka M, Satokawa S (2020) Rapid removal of resin from a unidirectional carbon fiber reinforced plastic laminate by a high-voltage electrical treatment. Sep Purif Technol 231:115885. https://doi.org/10.1016/j.seppur.2019.115885
44. Das M, Varughese S (2016) A novel sonochemical approach for enhanced recovery of carbon fiber from CFRP waste using mild acid–peroxide mixture. ACS Sustain Chem & Eng 4(4):2080–2087. https://doi.org/10.1021/acssuschemeng.5b01497
45. Ragupathi B, Bacher MF, Balle F (2023) First efforts on recovery of thermoplastic composites at low temperatures by power ultrasonics. Clean Mater 8:100186. https://doi.org/10.1016/j.clema.2023.100186
46. Perli G, et al (2023) Design for disassembly of composites and thermoset by using cleavable ionic liquid monomers as molecular building blocks. Compos Part B: Eng 264:110899. https://doi.org/10.1016/j.compositesb.2023.110899
47. Zhao J, Liu P, Yue J, Huan H, Bi G, Zhang L (2023) Recycling glass fibers from thermoset epoxy composites by in situ oxonium-type polyionic liquid formation and naphthalene-containing superplasticizer synthesis with the degradation solution of the epoxy resin. Compos Part B: Eng 254:110435. https://doi.org/10.1016/j.compositesb.2022.110435
48. Pathak GS, Hinge M, Otzen DE (2023) Transdisciplinary pragmatic melioration for the plastic life cycle: why the social, natural, and technical sciences should prioritize reducing harm. Sci Total Environ 895:165154. https://doi.org/10.1016/j.scitotenv.2023.165154
49. Nehls G (2022) European R&D project Bizente researches enzymatic technologies to recover thermosets [Online]. https://www.compositesworld.com/news/european-rd-project-bizente-researches-enzymatic-technologies-to-recover-thermoset-resins
50. Persico L, et al (2022) Recycling process of a basalt fiber-epoxy laminate by solvolysis: mechanical and optical tests. Fibers 10(6):55 [Online]. https://www.mdpi.com/2079-6439/10/6/55
51. Bhat T, Fortomaris D, Kandare E, Mouritz AP (2018) Properties of thermally recycled basalt fibres and basalt fibre composites. J Mater Sci 53(3):1933–1944. https://doi.org/10.1007/s10853-017-1672-7
52. Finnerty J, et al (2023) Effect of mechanical recycling on the mechanical properties of PLA-based natural fiber-reinforced composites. J Compos Sci 7(4):141 [Online]. https://www.mdpi.com/2504-477X/7/4/141

53. Hao J, et al (2024) Recycling and reuse of basalt fiber reinforced resin materials with artificial intelligence technology. J Electr Syst
54. Witik RA, Teuscher R, Michaud V, Ludwig C, Månson J-AE (2013) Carbon fibre reinforced composite waste: an environmental assessment of recycling, energy recovery and landfilling. Compos Part A: Appl Sci Manuf 49:89–99. https://doi.org/10.1016/j.compositesa.2013.02.009
55. Vo Dong PA, Azzaro-Pantel C, Cadene A-L (2018) Economic and environmental assessment of recovery and disposal pathways for CFRP waste management. Resour, Conserv Recycl 133:63–75. https://doi.org/10.1016/j.resconrec.2018.01.024
56. Song YS, Youn JR, Gutowski TG (2009) Life cycle energy analysis of fiber-reinforced composites. Compos Part A: Appl Sci Manuf 40(8):1257–1265. https://doi.org/10.1016/j.compositesa.2009.05.020
57. van Nunen JAEE, Zuidwijk RA (2004) E-enabled closed-loop supply chains. Calif Manage Rev 46(2):40–54. https://doi.org/10.2307/41166209
58. Farel R, Yannou B, Ghaffari A, Leroy Y (2013) A cost and benefit analysis of future end-of-life vehicle glazing recycling in France: a systematic approach. Resour, Conserv Recycl 74:54–65. https://doi.org/10.1016/j.resconrec.2013.02.013
59. Witik RA, Payet J, Michaud V, Ludwig C, Månson J-AE (2011) Assessing the life cycle costs and environmental performance of lightweight materials in automobile applications. Compos Part A: Appl Sci Manuf 42(11):1694–1709. https://doi.org/10.1016/j.compositesa.2011.07.024
60. Delogu M, Zanchi L, Maltese S, Bonoli A, Pierini M (2016) Environmental and economic life cycle assessment of a lightweight solution for an automotive component: a comparison between talc-filled and hollow glass microspheres-reinforced polymer composites. J Clean Prod 139:548–560. https://doi.org/10.1016/j.jclepro.2016.08.079
61. Meng F, McKechnie J, Pickering SJ (2018) An assessment of financial viability of recycled carbon fibre in automotive applications. Compos Part A: Appl Sci Manuf 109:207–220. https://doi.org/10.1016/j.compositesa.2018.03.011
62. Li H, Englund K (2017) Recycling of carbon fiber-reinforced thermoplastic composite wastes from the aerospace industry. J Compos Mater 51(9):1265–1273. https://doi.org/10.1177/0021998316671796
63. Curran R, Raghunathan S, Price M (2004) Review of aerospace engineering cost modelling: the genetic causal approach. Prog Aerosp Sci 40(8):487–534. https://doi.org/10.1016/j.paerosci.2004.10.001
64. Udavant S. Operating vs. Capital Expenses: What's the Difference? https://www.thebalancemoney.com/operating-vs-capital-expenses-what-s-the-difference-5219895
65. Schiller B (2001) Essentials of economics 4th ed. New York, NY, USA. McGraw-Hill/Irwin
66. Li X, Bai R, McKechnie J (2016) Environmental and financial performance of mechanical recycling of carbon fibre reinforced polymers and comparison with conventional disposal routes. J Clean Prod 127:451–460. https://doi.org/10.1016/j.jclepro.2016.03.139
67. Zhao X, Verhagen WJ, Curran R (2020) Disposal and recycle economic assessment for aircraft and engine end of life solution evaluation. Appl Sci 10(2):522

Recycled Synthetic Fibers Reinforced Composites (SFRC) and Their Applications

Usman Zubair

1 Introduction to Synthetic Fiber Reinforced Composites (SFRC)

Synthetic fibers of organic or polymeric origin such as aramid, polyethylene terephthalate (PET), polyacrylonitrile (PAN), polypropylene (PP), and polyethylene (PE) are occasionally used in SFRC, but they present some advantages over GFRC [1]. All types of fibers have specific characteristics that make them advantageous for particular applications over other fibers. For example, aramid fibers do better than glass fibers regarding specific tensile strength and stiffness. Similarly, aramid fibers and PET fibers composites exhibit better impact strength performance than glass fibers. These fibers exhibit lower densities than glass fibers, making them pertinent for lightweight applications. PET fibers reinforced polypropylene exhibit better strength and modulus than that of sisal-reinforced polypropylene composite. PET fiber polypropylene exhibited a higher value of maximum strain than that of aramid fiber, glass fiber, and sisal fibers-based polypropylene composite. These fibers are also found more suitable for their application in non-polymeric matrices such as concrete and gypsum. Table 1 presents the density, tensile strength, tensile modulus, and strain percentage of various synthetic fibers against glass fibers, carbon fibers, and stainless steel.

U. Zubair (✉)
Department of Materials, School of Engineering and Technology, National Textile University, Faisalabad, Pakistan
e-mail: usman.zubair@ntu.edu.pk

H. Jamshaid and U. Zubair (eds.), *Sustainable Recycling of Fiber Reinforced Polymer Composites*, SDGs and Textiles, https://doi.org/10.1007/978-981-96-9536-2_6

Table 1 Comparison of the density and mechanical properties of synthetic fibers against E-glass, carbon, and stainless-steel fibers

Fiber's type	Density (g/cm^3)	Tensile strength (GPa)	Tensile modulus (GPa)	Strain (%)
UHMWPE fibers	0.97	2.5–3.9	97–133	2.8–4.5
Aramid fibers	1.44	1.4–3.6	62–190	1–4.4
Arylate fibers	1.4	2.5–3.4	75–110	4
Polyamide 66	1.14	0.91	9.57	37
Polyester (PET)	1.4	0.7–0.9	12–17	7–37
E-glass fibers	2.55	2	90	1.8–5.4
Carbon fibers	1.8	3.5–7.5	230–540	0.7–2
Stainless steel	7.9	2	210	1.4

1.1 Polyolefins Fibers-Based SFRC

Polyolefin fibers are those in which long chain polymers at least 85% by mass are composed of ethene, propene, and other olefins. Fiber reinforced polymer composites are strongly making their way into vast fields of application owing to durability, cost-effectiveness, weight reduction, and high specific strength. But the recyclability of the FRPC is challenging and pose a threat to their sustainable use. This is the very same reason that polyolefins fibers are emerging as a perspective solution because of their reuse and recyclability [2]. The major use of polyolefin fibers is in concrete reinforcement that improves the strength of concrete. Ductile nature of polyolefin fibers improves the impact strength of the composites. Owing to excellent impact properties, polyolefin fiber composites are used in ballistic armor. However, polyolefins are now also preferred as thermoplastic matrices in combination with natural fibers reinforcers. Ultra-high molecular polyethylene fiber (UHMWPE) is considered as a high-performance fiber is utilized as reinforcement in epoxy resin. Potassium dichromate-treated UHWMPE fiber epoxy composites exhibited a higher impact strength than that of pure epoxy composite and untreated fiber composite owing to the formation of (-OH) groups [3]. Similarly, there is an improvement in the modulus and bending strength of the composite.

1.2 Polyester Fibers-Based SFRC

Polyester fibers have been extensively used as a skeleton material in rubber composites to enhance their strength and durability. However, interfacial adhesion between rubber and polyester fibers is weak owing to the chemical inertness of PET fibers [4]. Polyester fiber cords are impregnated with resorcinol–formaldehyde-latex (RFL) to enhance the adhesion between the two materials. Resorcinol is injurious to human health, so different resins have been investigated. Meng et al. have investigated cresol-formaldehyde latex (CFL) that is simple and less toxic owing to the use of m-Cresol [5]. They have reported remarkable enhancement in adhesion using CFL reaching the peeling force value and the H pull-out force of treated PET fiber/rubber composites to 7.3 N/piece and 56.8 N, respectively.

1.3 Polyamide Fibers-Based SFRC

Polyamide fibers bear amide linkages through their polymeric structure. Aramid fibers have been extensively utilized in the construction of composites for flame-resistant applications, high-temperature protective gears, body armor, thermoplastic pipes, and rubber reinforcement. Para-aramid fibers are rigid aromatic polyamide, poly para-phenylene terephthalamide (PPTA), that are axially aligned and radially bonded by hydrogen bonds that grant them with very high tenacity and modulus. Aramid fibers are marked for their exceptional strength that is six times of steel. Aramid fibers account for about 7% of the market share. Reinforcement of aramid fibers into epoxy or polyester resins yields the composite with extraordinary tensile strength (relatively poor compressive strength). These composites exhibit a high strength-to-weight ratio, abrasion resistance, non-conductive, and heat resistance. Para-aramid fibers are commercially produced by DuPont, Teijin and Yantai Tayho Advanced Materials Co. Ltd under the brand names Kevlar®, Tawron®, Technora®, and Taparan®. The complicated processing of poly para-phenyleneterephthalamide into fibers makes them relatively costly fibers. This holdup makes it essential to recycle aramid fibers for their multiple uses in addition to reaching a sustainable economy.

Aramid fiber composites have been proven a hallmark in protective antiballistic applications owing to the ability of aramid fibers to absorb enormous amounts of impact energy arising from modern ammunition and weapons [6]. These properties arise from the high modulus of the aramid fibers that can be credited to hydrogen bonding, compact molecular structure, high crystallinity, and high density. Epoxy-based thermoset resins are widely used in the construction of antiballistic composites. There are various strategies to enhance the antiballistic properties of these composites including the inclusion of shear-thickening fluids, layer-by-layer assembly, yarn pullout, and surface and nanoparticle functionalization of aramid fibers. Surface modifications of aramid fibers have been realized using various nano moieties such as silica, ZnO, graphene, carbon nanotubes, and silk fibroin. The modification of fibers prevents the fibers from buckling and contributes toward performance enhancement. Recently, there has been an emphasis on the use of thermoplastic resin that allows the convenient recovery and recycling of raw materials.

Some unique approaches have also been investigated in which the thermoset resin is replaced with a novel resin system with a unique chemistry that allows the convenient recyclability of the composites. In a piece of research work, a novel resin has been developed with a dynamic covalent adaptive network using Diels–Alder chemistry to replace epoxy/aramid composite [7]. The as-fabricated composite has demonstrated comparable properties to that of epoxy composites. Still, it can easily be recycled into neat aramid reinforcement by dissolution in some solvents credited to Diels–Alder chemistry.

2 Upcycling/Recycling of Plastic and Textile Waste into Recycled SFRC

The depletion of fossil fuel resources and the non-recycling of synthetic fibers are becoming a future global crisis and pose a severe threat to the ecological system due to the non-compostability of plastic and synthetic fibers. The end-of-life scrap comes from various sources including but not limited to textile manufacturing waste such as loose fiber, chopped fibers, shredded fibers from textile fabric waste, cured and semi-cured prepreg waste, and fully cured composite structures from various industries including aerospace, automobile, pressure vessels, windmill blades, and boat hulls. Lin et al. have demonstrated the use of recycled high-strength polyester and aramid fibers recovered from the selvage wastes of woven fabrics [8]. They have constructed sandwich structure composites where the surface layer is yielded from recycled aramid, nylon, and low-melting polyester fibers using a non-woven process. The interlayer is composed of nonwoven high-strength polyester. The process of needle punching has been deployed to achieve nonwoven structures. The as-obtained sandwich structure composite from recycled fibers demonstrated good strength, high impact resistance, and high energy absorption. This is an example of upcycling that shows how industrial waste can be transformed into valuable products such as protective gear. This approach reduces waste and environmental pollution and results in products of high-end value.

Poulakis et al. have demonstrated the potential reuse of modified aramid fibers obtained from the recycling process [9]. They have successfully reclaimed aramid fibers and matrix from ionomer-based composites through dissolution, reprecipitation, and filtration techniques. As-recovered fibers were embedded into fresh ionomers for the fabrication of second-generation composites. Interestingly, the composites from recycled fibers exhibited better tensile properties that can be credited to the modified surface of recycled fibers, better dispersion, and redistribution into the ionomer matrix. Vaidya et al. [10] have presented the wet-compression molding technique as a scalable approach to produce large composite panels from mixed shredded waste coming from multiple sources. The composite panels are potential candidates for use in truck bodies and flooring panels of intermodal shipping containers. The shredded waste came from various waste streams of post-industrial and end-of-life scraps of textile and composite manufacturing facilities. The properties of the as-fabricated panels are independent of the waste stream and highly acceptable for the mentioned applications. Both synthetic fiber reinforcements and resin matrices are derived from crude oil through the extraction of raw materials

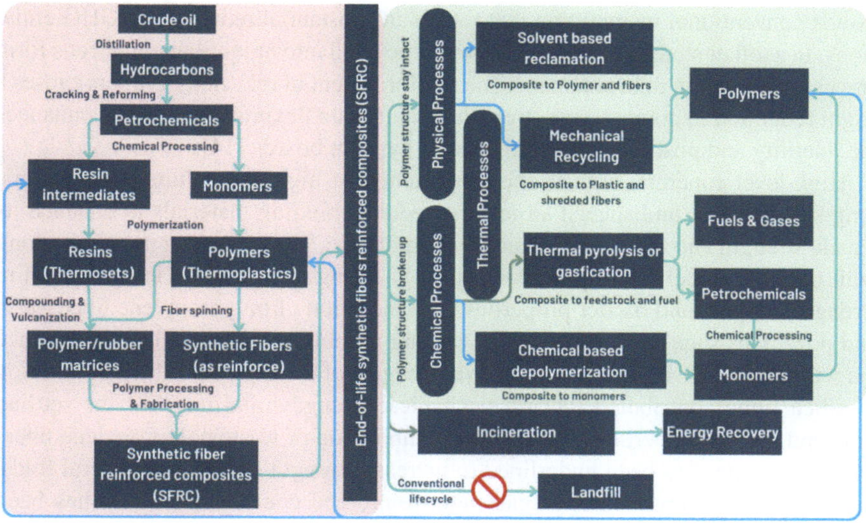

Fig. 1 Extended value chain (life cycle) for bringing the circular economy in SFRC

from petrochemicals. At the end of their life, SFRC can be recycled further through mechanical, thermal, and chemical recycling, in contrast to conventional disposal approaches such as landfills and incineration (Fig. 1). This extended value chain, in the form of recycling, allows us to close the loop to bring the circular economy.

2.1 Polyolefins Fibers-Based Recycled SFRC

Polyolefin fibers such as polyethylene (PE) or polypropylene (PP) fibers have been extensively used in the manufacturing of technical textiles including Sporttech, Medtech, Geotech, Agrotech, and others. For example, in the recent past during COVID-19, the use of face masks has soared to an all-time high in human history with a consumption of 129 billion masks per month globally [11]. These masks are constructed of polypropylene melt-blown nonwoven fabrics in three-, four-, or five-layer constructions. They emerged as one of the biggest sources of plastic pollution and only 1% of improperly disposed masks will result in 30–40 tons of plastic waste that goes into the environment [12]. In a study, end-of-life sports pitches (artificial turf) that are made of PE and PP have been recycled back to fibers and studied for their use in concrete composites instead of virgin fibers [13]. It is found that recycled fibers induce a ductile post-cracking regime in bending and the composite exhibited similar toughness as that of virgin fibers composite. Through LCA, it is established that there is a considerable reduction in ecological footprint with a two-prong benefit to the environment. First is the reduction in the need for virgin plastic and second, it avoids conventional incineration that results in substantial reduction in GHG emissions. In another study, PE fishing nets were recycled into monofilament fibrous form and used in gypsum reinforcement. The reinforcement of recycled PE fibers causes a slight reduction in compressive strength and ultimate flexural strength but enhances the ductility and post-crack performance of gypsum boards [14].

High-level concrete, used in the construction of high-rise columns and towers, long-span bridges and special structures, requires binding materials to enhance its tensile strength and reduce its sensitivity to moisture. These binding materials along with fiber reinforcers are a promising strategy owing to significant improvement in strength, fatigue, and barrier properties. The inclusion of polypropylene fiber waste and pozzolans such as silica fumes in varying proportion has significantly enhance the workability, tensile, and compressive strengths of the concrete [15]. The addition of silica fume contributed toward mechanical strengths and improve the settling time and barrier properties. The optimal combination of waste polypropylene fibers and silica fumes has been underlined to increase strength of concrete several folds. In other study, prepacked aggregate fiber-reinforced concrete (PFARC) has been investigated for its long-term strength properties, thermal properties, and resistance to acids and sulfates [16]. In this study, waste polypropylene fibers and pozzolan such as palm oil fuel ash (POFA) were mixed in the concrete. The combination provided with a denser microstructure that lowers the heat transfer to middle of concrete and controlled loss in strength in acid and sulfate environment. PP fibers enhance the long-term strength of the as-constructed PFARC.

2.2 Polyester Fibers-Based Recycled SFRC

PET has been extensively used for the manufacturing of beverage packaging bottles and synthetic textiles. It can be used to reinforce matrices such as low-density polyethylene (LDPE) to form microfibrillar-reinforced polymer–polymer composites (MFC). MFC are achieved through reinforcing microfibrils of higher melting polymer into low melting temperature isotropic matrix such as PET fibrils in LDPE matrix [17]. Recycled PET from bottles is melt blended into LDPE matrix in 30/70 and 50/50 blend ratio by weight and injection molded after extrusion, drawing, and pelletizing. The drawing causes the fibrils to get oriented within an isotropic matrix confirmed by X-ray scattering. In the injection molded sample, the microfibrils are randomly distributed but cause the samples to exhibit impressive mechanical properties. A 10-time increase in elastic modulus was reported in comparison with clean LDPE samples, three times higher than glass sphere reinforced samples and equivalent to 30% GF reinforced fibers. The impact strength of the as-prepared samples increased by 50% to that of neat LDPE. This approach offers an alternative to reusing PET and other polymers instead of conventional recycling approaches.

The waste polyester fibers obtained from silk wadding polyester quilt are extrusion blended within recycled polyethylene plastic. Waste polyester fibers served as reinforcer in polyethylene plastic and alter the microstructure of the plastic products.

This reinforcement improved the thermal stability and mechanical properties such as tensile strength, and flexural and loss moduli. The maximum level of mechanical properties is reached at 20 wt% reinforcement [18]. In other investigation, Miao et al. have proposed the inclusion of waste polyester fibers into polypropylene to realize composite boards for furniture applications [19]. The polyester fibers waste was collected from the waste polyester clothing. The induction of polyester fiber improved the mechanical, thermal, and insulation properties without altering the chemical and crystalline structure, and hydrophilicity of polypropylene. As-fabricated furniture panels exhibited the superior performance than that of available furniture materials in the market. In another study, a fire-safety composite felt was fabricated through a cost-effective opening, combing, and needle-punching technique using recycled PET waste and inherently flame-retardant Ca-doped alginate fibers. The as-prepared composite fibers and felts can be effectively used in fire-proofing construction material as filler and insulating material [20].

The end-of-life tires are recycled by mechanical grounding that provides us with recycled tire fibers (RTF) and ground tire rubber as particles. The nature of the reclaimed RTF depends on the tire cord which is usually polyester, while obtained ground rubber particles are of varying sizes in micron range. In an investigation, polyester RTF mixed with ground tire rubber particles were used as fillers in LLDPE composite through extrusion and injection molding techniques. Styrene–ethylene–butylene–styrene-grafted maleic anhydride was used as a compatibilizer up to 10% to enhance the adhesion among the materials. The variations in RTF content, extruder screw speed, and temperature profiles were investigated both for injection and extrusion. It is found the optimal properties were realized at maximum RTF content (i.e., 50%), extrusion speed 250 rpm, and lowest temperature profile with compatibilizer [21, 22]. In another work, Kazemi et al. have segregated the mixture of RTF and grounded rubber particles into clean fibers and residue grounded rubber [23]. The composites from both components were prepared at different weight percentages using recycled LDPE as a matrix. They have observed that composite with clean fibers at 30 weight percent offers an increase of 15% in tensile strength, 192% in tensile modulus, and 142% in flexural modulus. While the inclusion of rubber particles at 30% in recycled LDPE reduces both tensile strength and flexural modulus by 15%. Both compression and injection molding were chosen for the preparation of composites. It is important to clean the tire-grounded waste before its use in composite materials.

Bocci et al. have reported the use of reclaimed fibers from end-of-life tires in hot-mix asphalt [24]. Rubber crumbs, steel wires, and textile fibers are among the waste reclaimed from end-of-life tires. Textile fibers account for 10% of the total waste obtained, which ends up in landfills or is incinerated. This study mixes reclaimed fibers with hot-mix asphalt to improve their mechanical performance. Although there is no nominal improvement in strength and stiffness, the fatigue resistance has drastically improved. The inclusion of waste fibers helps to sew the microcrack edges and inhibit the evolution of them in macrocrack. This strategy allows the use of fibers from tire waste to be effectively used to enhance the mechanical performance of asphalt.

Teijin Frontier Co., Ltd., claimed the development of the first tire cord fabricated out of an eco-friendly adhesive and recycled polyester for rubber composites for the tire industry. Conventionally, resorcinol formaldehyde (RF) latex is used as an adhesive to increase the interfacial bonding between rubber and polyester cord. RF is considered harmful to life and the environment. The company claimed to use RF-free adhesive and low-impact chemically recycled polyester for the fabrication of tire cords. The company claims to reach its annual target of 200,000 tons by 2030 after its successful run in the same year [25]. The tire cord is supposed to have the same strength and performance properties but lower the CO_2 emissions and environmental impact. This development highlights the involvement of industries in recycling ventures for a better future and creating a social impact.

2.3 Polyamide Fibers-Based Recycled SFRC

Regenerated aramid short fibers, coming from the production scrap of aramid yarn production facility, were utilized for the reinforcement of commercial polyamide 12 [26]. The reinforcement content varies from 10 to 30% by weight to underline the effect of reinforcement on mechanical, thermomechanical, and dynamic properties. The thermal stability of polyamide fiber composites slightly improved and the softening temperature was raised by 10–20 °C, while the isothermal creep was substantially removed. The deformation at break and coefficient of linear thermal expansion is also reduced. The stiffness and embrittlement of polyamide composites are increased due to the stabilizing effect of incorporated recycled fibers. The change in properties is gradual and highly dependent on the percentage content of fibers in the composites. In a study, the recycling of fibrous fractions from epoxy-based thermoset composites and the use of recovered short fibers in the fabrication of thermoplastic composites [27]. Aramid fiber composites were investigated in addition to carbon and glass fiber-reinforced composites. The reclaimed short fibers-based composites exhibited comparable properties to the similar to that of short virgin fibers composites. The fibers are recovered from thermoset composites through mechanical grinding and sifting.

The waste aramid fibers from industrial waste can be mechanically reinforced in polyamide 66 matrix [28]. The aramid fibers have to be surface functionalized before the inclusion to enhance the interfacial adhesion. For this purpose, hexane washing of the fibers has been performed followed by NaOH solution treatment at varying conditions. It has been found that 10 wt% addition of treated aramid fibers has positive effect both on interfacial adhesion and mechanical properties. The realization of these composites would open new avenues for various engineering applications. In another investigation, Chaung et al. studied the inclusion of aramid staple fiber waste from the textile fabric selvages into low melting polyester [29]. Aramid fibers were retrieved by mechanical opening were transformed into nonwoven. These nonwovens were further shaped into composite boards by impregnating in low melting polyester. The boards with 90 wt% recycled aramid fibers offered an optimal combination of both mechanical strength and static puncture resistance. This strategy of mechanical shearing and mixing of aramid fibers offers a cheap solution for high-performance composite boards. Ma et al. have invented a viable top-down approach to synthesizing aramid nanofibers and cellulose nanofibers from waste resources [30]. These nanofibers were introduced into carbon fiber/phenolic resin composites as nanofillers. The introduction of these nanofibers has significantly enhanced the tensile strength and modulus, flexural strength and modulus, shear strength, and elastic modulus of the carbon fiber/phenolic composites. This strategy underlines the right exploitation of waste resources to high-end advanced functional materials.

3 Recycling of SFRC: Levels and Routes

A circular economy in composites can be reached by enabling recycling technologies. Due to composite materials' integrated structure, it becomes inherently difficult to separate the individual components. Recycled materials incur additional monetary and energy costs in comparison to virgin materials. Recycled materials are always at lower levels of quality due to the deterioration from recycling processes and usage wear and tear. Being of poor-quality materials, they can only be used in fewer applications. The used materials can also be converted into their feedstock elements. Recycling SFRC is relatively challenging compared to glass or carbon fibers owing to the chemical sensitivity of synthetic fibers. The glass and carbon fibers are chemically inert and can be reclaimed conveniently by melting the thermoplastic resin or degrading the thermoset resin. Synthetic fibers are vulnerable to heat and chemical environments, for example, ester bonds are susceptible to hydrolysis in the presence of moisture and heat. The recycling of synthetic fibers can be ranked as primary, secondary, tertiary, and quaternary stages following the 9R framework of waste management. This framework offers a comprehensive set of strategies for waste management and the circular economy, including removal, reduction, resourcing, reusing, recycling, rethink, refurbish, repair, repurpose, recovery, and return. The primary level of recycling includes reprocessing, the secondary level includes mechanical and dissolution recycling, the tertiary level involves chemical depolymerization, pyrolysis, and gasification, while quaternary level, as a last resort, includes incineration for energy recovery (Fig. 2).

4 Waste Prevention and Reuse

In the hierarchy of sustainable practices, removal, reduction, and resource efficiency are at the top of the pyramid. These steps include the measures taken before the production of waste such as the substitution of hazardous materials with non-hazardous ones, reducing the resources required to produce composite products,

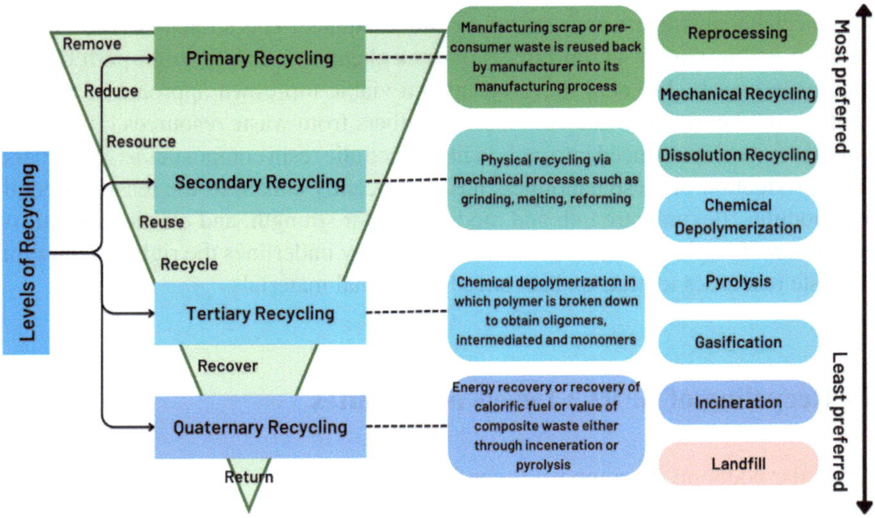

Fig. 2 Levels and routes of composite waste recycling and their relation

enhancing the efficacy of the involved processes, and extending the life of the products. Another option that appears before recycling is the reuse of composite products either for a similar application or an application in a different field. The process of reuse does not involve any thermal, mechanical, or chemical transformation of the materials used in the product.

4.1 Waste Prevention/Eco-Designing

Waste prevention can be addressed at multiple avenues, starting from the eco-designing of the products. Eco-designing is the conscious designing of the product that either extends the life cycle of the products, allows the convenient recycling of the products, or both. Eco-design reduces the waste from the end-of-life cycle products. Garfias et al. have designed aramid fibers-based chemically recyclable epoxy composites by surface modification of aramid fibers. They have altered the surface of aramid fibers by self-polymerization with dopamine followed by condensation with a silane coupling agent. Such surface modification enhances the interfacial bonding between aramid fibers and resin and allows the attainment of circularity. They claimed the removal of 82% of epoxy resin by mild solvolysis in ethylene glycol for the effective reclamation of surface-modified aramid fibers [31]. There is a shift toward thermoplastic composites in contrast to thermoset composites, as thermoplastic matrices can be melted again and again and can be recycled easily. However, thermoset resin-based composites are still in popularity. The introduction of vitrimers allowed the processability and recyclability of thermosets by inducing covalent adaptable networks. Covalent adaptable networks allow the vitrimer to retain solvent resistance and mechanical stability as well as permit fluidity of the matrix on heating [32]. In the previous section, various studies have been pinpointed that are linked to the inclusion of waste high-performance synthetic fibers into common thermoplastics such as polyethylene, polypropylene, and nylon. The introduction of these fibers improves the mechanical strength, thermal stability, and barrier properties of common plastic components. This approach allows the use of waste to reduce the plastic waste by reducing the material required for the construction of similar products but with an extended lifecycle.

4.2 *Waste Reuse*

Thermoplastic composites can be remolded by thermomechanical action. The use of aramid fiber waste, arising from the production scrap of aramid yarn production, in the reinforcement of commercial polyamide 12 is an example of the reuse of the waste material into SFRC [26]. In a process, the thermoplastic composite is fragmented into large pieces, then laid on the conveyor belt with uniform distribution, and restored into large panels by performing sheet compression molding [33]. Other processes involve the conversion of grounded composite waste into dough through shearing and heating, and finally, compression molding of the dough to achieve the products of the required shape. Discarded waste tires are now considered a source of valuable raw material that can replace virgin raw materials for the construction of both thermoplastic and thermoset composites [34]. Waste tire rubber is used as a filler in the form of granules for the construction of low-cost and environmentally sustainable composites.

5 Fabrication and Processing Technologies for Recycled SFRC

The literature shows that several plastic processing technologies are used for recycled SFRC. The most common technologies implemented for the reuse of SFRC are extrusion, injection molding, compression molding, resin transfer molding, and sheet molding. Figure 3 presents the schematic of the mentioned techniques. These processes are used in combination to fabricate recycled composites. For example, thermoplastic composites are mechanically shredded into small pieces that are mixed under heat and shear in an extruder to make a dough, which is compressed into the required shape on a compression molding apparatus. In another approach, the grounded residue of tires is mixed with the resin and compression molded into other low-strength products, such as tennis balls. In a study, the shredded scrap from different waste streams was intermixed and subjected to wet compression molding

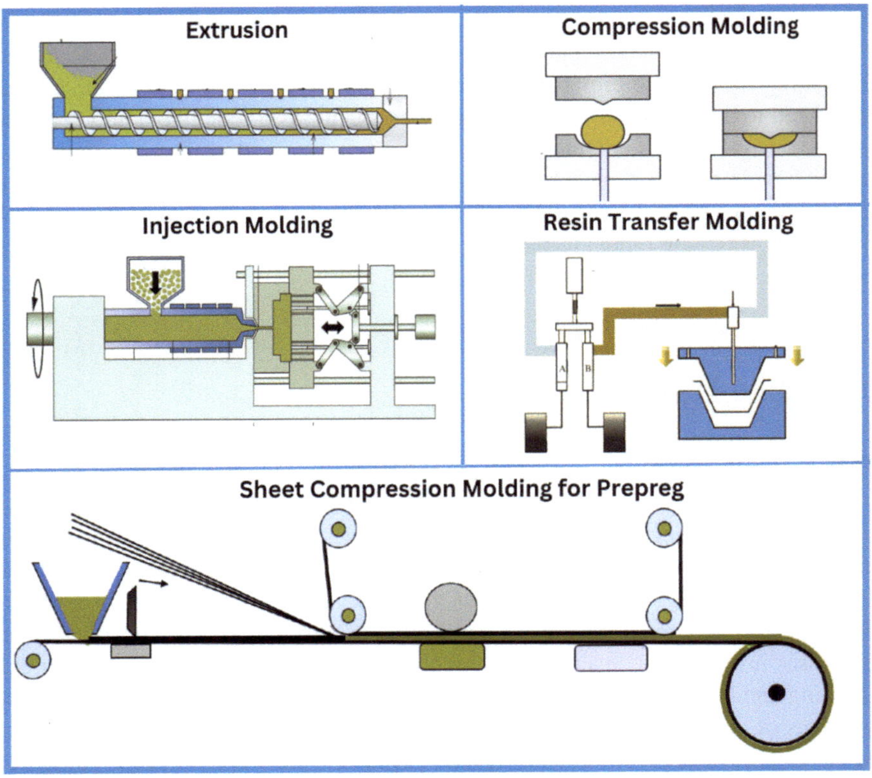

Fig. 3 Representation of different recycled SFRC composite fabrication technologies

to fabricate composite panels for truck bodies and containers' flooring panels. Wet compression molding was found to be a practical and scalable approach for producing composite panels [10]. The waste polyethylene plastic is first blended with waste polyester fibers coming from polyester silk wadding quilts using extrusion molding. Then the composite is shaped through injection molding to realize the plastic with better tensile strength, loss modulus, flexural modulus, and thermal resistance [18].

6 Physical Recycling

Physical recycling involves the recovery of composite components without any significant chemical changes. It changes the physical state of the materials either through mechanical recycling or dissolution recycling.

6.1 Mechanical Recycling

Mechanical recycling involves the size reduction of large composite scraps into 'small elements that can be further separated into their components. Mechanical recycling could be carried out through either mechanical grinding or high-voltage fragmentation. Composite parts are preliminarily shredded into pieces of varying sizes from 50 to 100 mm on slow speed cutting or grinding machines. Then the process of grinding is carried out on high-speed cutting or hammering mills as depicted in Fig. 4 to reduce the size of the shredded pieces into fragments of less than 10 mm to 50 μm. These fragments are separated into resin-rich powder and fibrous fragments through cyclones and sieves. The estimated energy demand for the grinding process is between 0.1 and 4.8 MJ/kg depending on the machines and scale of the process. Resin powder can be used as filler in the first place and offers lower weight addition towards newly designed parts. Mechanical recycling appears cost-effective and eco-friendly, although it deteriorates the fiber properties such as fiber's length and strength. The fiber length is important if the textile fiber waste needs to return to yarns and fabrics. But, they can be effectively used in technical non-wovens and composites. For example, recycled PET fibers were obtained through economical opening, combing, and needle-punching techniques in a work. These fibers were composited with Ca-alginate fibers to realize a composite felt with excellent flame retardancy. These felts can be used as fire-proofing filler and fire-safety composites for the construction industry and buildings [20]. High voltage fragmentation (HVF) is an alternative electromechanical approach that uses electricity to separate the fiber reinforcement from resin matrices. This process retains the fibers in their original shape with minimal resin residual content, but the process is energy-intensive and requires 17.1 to 89.1 MJ/kg of energy [35]. In the process, high-voltage pulses are passed through the composite scraps, and placed in a dielectric liquid usually water. HVF is beneficial in the process of recycling composites that are constructed from high-end fibers. This technology is under development and reached a technological readiness level (TRL) of 6–7.

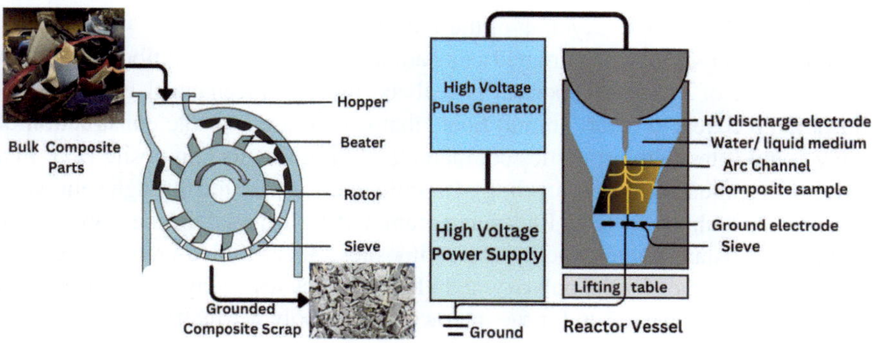

Fig. 4 Mechanical recycling of waste composites; mechanical grinding (left), high voltage pulse fragmentation (right)

Tires are essentially rubber composites involving the use of synthetic fiber cords. These tire cords are made of polyester and polyamide and coated with an adhesive to enhance the interfacial bonding between rubber and cord. Owing to vulcanization and crosslinking in rubber structures, mechanical recycling is used to separate rubber crumbs from fibers. W. S. Anthony has reported a new machine that can effectively separate the waste fluff produced by tire recycling plants into marketable products such as crumb rubber and fiber [36]. Between the two lab-developed prototype machines, one was tested industrially at a commercial recycling facility and found to yield 70% rubber crumb with size $\leq$ 300-micron, 13% clean fiber, 9% partially cleaned fiber, 5% large pieces of un-ground rubber, and 3% cyclone waste. On changing the raw materials to high variability, the ranges change considerably. However, the economic evaluation of three tire-recycling facilities with the capacity to process 2 to 5 million tires annually indicates an income growth from \$0.67 to \$1.94 million per annum. The profits depend upon the quantity of tires to be processed and the cost of dumping the material into landfills.

The fibrillation of aramid fibers in composite using a vibrating ball mill is a mechanical recycling approach [37] operating conditions, such as processing time, can control the morphology of the fibers. The fibrils are peeled off from aramid fibers that exhibit poor dispersibility and multiscale size. The recycled aramid fibrils are good enough for further use but not as good as virgin fibers, and the process is highly energy-demanding.

Mechanical recycling although is preferable option for the recycling of the composites but the process of mechanical recycling deteriorates the fibers physical properties especially the mechanical strength for which these fibers are valued. Garfias et al. have proposed the surface functionalization of short-cut aramid fibers for the continent mechanical recycling of the fibers from the epoxy composite [38]. On one hand the use of covalent adaptable network enables the reversible bonds in epoxy composite during recycling. Epoxy-based covalent network adaptable matrix is achieved through blending of citric acid and tri-epoxy resin. While on the other end, the polydopamine-silane functionalization of aramid fibers ensures the fibers' mechanical strength retention by three times to that of untreated fibers. This approach not only enhances the composite performance but also offers more sustainable solutions for recycling of thermoset composites.

Epoxy composites are constructed using aramid, carbon, and glass fibers. Mechanical recycling of epoxy composites is realized through mechanical grinding that allows the recovery of short aramid fibers that can be used for the construction of epoxy composites with properties comparable to virgin fibers composite [27]. In a study, environment-friendly recycling of aramid fibers is realized through solid-state shear milling technology [39]. Recovered aramid fibers are used for the construction of friction-resistant polyoxymethylene composites. Shear-milled aramid fibers act as a friction modifier for the polyoxymethylene composites owing to micro size and presence of functional group for the interaction with polymer matrix.

6.2 Dissolution Recycling

Solvent-based reclamation of plastics and composites, also termed dissolution recycling, involves the physical recovery of the polymers using the solvent system. It does not involve the chemical degradation of plastic as is the case with solvolysis. It is a physical process in which the targeted polymer from the mixed composite scrap is dissolved in a suitable solvent. The chemical structure of the polymer remains intact and the other components in the composite waste such as fibers, additives, fillers, and undesired polymer stay undissolved. The undissolved components are separated from the solution of the targeted polymer. The cleaned polymer is separated from the solution by either adding antisolvent or heating the mixture. The addition of antisolvent would precipitate the polymer. The heating of the mixture would also allow the removal and recovery of solvent for closed-loop processing as shown in Fig. 5. In a study, Zhou et al. have designed an aramid fabric reinforced composite where a high-performing dynamic covalent adaptive network based on Diels–Alder chemistry was used as a matrix and aramid fabric as reinforcement [7]. They have exhibited comparable properties to conventional epoxy/aramid composite. But the matrix can be easily dissolved in solvent as aramid fabric can be recovered with the properties of virgin aramid fabric. Owing to the matrix covalent adaptive network with Diels–Alder chemistry, these composites can be remolded to the target shapes.

7 Chemical Recycling

Chemical recycling of synthetic fibers and resin involves depolymerizing polymeric chains into monomeric units. In the process, the matrix can be dissolved without affecting the properties of fibers. There are multiple possibilities of chemical recycling including hydrolysis, solvolysis, glycolysis, acetolysis, carbolysis, aminolysis,

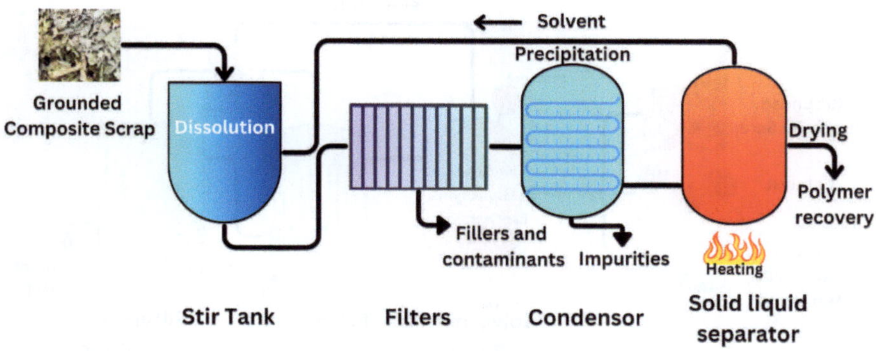

Fig. 5 Schematic of dissolution recycling of composites

and enzymatic depolymerization. This categorization is based on the type of solvent involved in the process.

7.1 Hydrolysis

Hydrolysis of the polymer is performed under high temperature and pressure in the presence of some catalyst using water as a solvent. Hydrolysis can be further differentiated as acid, alkaline, and neutral hydrolysis. Figure 6 depicts the equipment used for the hydrolysis of composite scrap. In the process, the reactor is fed with some solvent (water), catalyst, and waste composite under heat and pressure. The hydrolyzed products along with the solvent passed through filters to separate the impurities, fibers, fillers, and unhydrolyzed waste. The impurities-free feed is passed through the distillator to separate the solvent, monomers, and byproducts. The recovered solvent is fed back to the reactor to close the loop. Recently hydrolysis of unsaturated and vinyl polyester resins is promoted using aprotic solvents. In a study, An et al. demonstrated the complete conversion of waste unsaturated polyester resin (UPR) into valuable feedstock using tetrahydrofuran (THF) as an aprotic solvent [40]. The use of the THF-water system was also found helpful in the convenient separation and recovery of feedstock. THF facilitates the fragmentation of resin by high diffusivity and ensures greater mass transfer. Water in lower quantities accelerates the degradation of the partially hydrolyzed resin. In another study, the hydrolysis of epoxy resin is enhanced using permeability-enhanced surfactants such as ammonium isooctyl sulfate (IAS) along with water [41]. The presence of IAS by 30 wt.% in water catalyzes the degradation of epoxy resin by enhancing the permeability of the system. Epoxy resin is completely hydrolyzed into oligomers with intact backbone chains at only 160 °C. They have signified this process as a transfer from surface erosion to volumetric degradation.

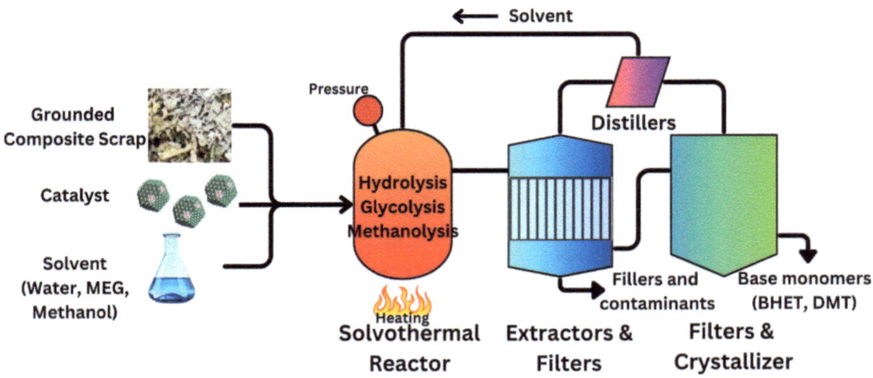

Fig. 6 Schematic of the process of solvolysis of polymer composites

7.2 Solvolysis

Polymeric resins could dissolve into unique solvent systems by the degradation of the links among the polymeric chains. Solvolysis could be realized using many solvents with different catalysts at varying temperatures and pressure conditions. Based on the used solvent, these are further categorized as glycolysis (using glycols), alcoholysis (using alcohol), carbolysis (using carboxylic acids), etc. Figure 6 presents a typical apparatus used in the process of solvolysis of composite waste. Yuyan et al. have effectively demonstrated the recovery of aramid fibers from aramid/epoxy composites through chemical degradation using solvents. In the process of solvolysis, 1,2,3,4-Tetrahydronaphthalene has been used as the main solvent. The epoxy resin was effectively decomposed using this solvent under optimal conditions and aramid fibers were successfully reclaimed [42]. The optimal reaction temperature was found to be 310 °C and the reaction time was 1 h. The reclaimed fibers had shown only an 8% reduction in strength which makes this process ideal for the recovery of high-strength aramid fibers.

Garfias et al. have carried out the effective chemical recycling of epoxy/aramid composites by exploiting acidic and basic hydrolysis in 1M solution of HCl and NaOH, respectively, and mild solvolysis using ethylene glycol as a solvent and zinc acetate as catalyst through a transesterification exchange reaction [31]. The presence of a high content of ester linkages allows the effective solvolysis of thermoset resin. The effective removal of thermoset resin is assisted by surface modification of aramid fibers with dopamine and a silane coupling agent. The reclaimed aramid fiber meshes retain their structural integrity and mechanical properties after recovery from composites.

Kirshanov et al. have reported the concurrent homogenous glycolysis of polyester cord and devulcanization of rubber from waste tires [43]. The devulcanization of 46% is realized using the swelling method for tire rubber using a softener as a reaction promoter, and the highest yield of bis(2-hydroxyethyl) terephthalate has also been reached in this process. Aramid fibers are extensively used in composite form for both civil and military applications. The mechanical recycling of aramid waste composites provides fibers with much inferior strength that does not allow their high-end usage. However, the chemical recycling of aramid fibers into nanofibers offers a much better solution to this problem and provides better functional materials [44]. This upcycling of aramid is not viable for industrial-scale application owing to the inefficiency of the processes. Chen et al. have proposed a novel deprotonation system comprised of KOH and dimethyl sulfoxide to develop nanofibers of well-defined and uniform structure. As per their claim, the proposed system is rapid and offers better throughput than other systems in the literature [45]. The realization of diverse microstructures also offers better recycling opportunities.

8 Thermal Recycling

SFRC is constructed from petroleum, hence its waste has recoverable calorific value and is a potential source of hydrocarbons. Thermal recycling involves applying heat energy to crack long-chain polymers into short hydrocarbons (aliphatic or aromatics) and transforming them into various forms of fuel (solid, liquid, or gases). Thermal recycling could be used either for the recovery of hydrocarbons or base chemicals as fuel or monomers or for direct energy reclamation from waste burning under controlled environments. In the former, pyrolysis, gasification, cold plasma pyrolysis, and plasma arc gasification are used, while combustion, incineration, and co-incineration are exploited for the latter. The main difference in these processes is the amount of oxygen under which the processes have been carried out. Figure 7 depicts the attained process temperature, oxygen requirement by the process, and the resultant outputs from different thermal recycling processes. Pyrolysis and gasification have been discussed under this heading, while combustion, incineration, and co-incineration will be addressed in the energy recovery section.

8.1 *Pyrolysis*

Pyrolysis is the most studied thermal process in which composites are burnt in the absence of oxygen or at a very low concentration of oxygen. The presence of a small amount of oxygen curtails the char formation. It allows the recovery of fillers, inserts, and fibers. Fiber recovery is only possible in the case of inorganic fiber composites. SFRCs are wholly made of petrochemicals, unlike other composites such as natural fiber, carbon fiber, inorganic fiber, and glass fiber composites. SFRCs can become a great source of hydrocarbons to replenish the feedstock and recovery of fuels. By changing the process parameters, SFRC can be transformed into solid, liquid, and gas fuels in the form of olefins, aliphatic, and aromatics. The pyrolysis is performed at the temperature range 400 to 700 °C depending on the resin to be pyrolyzed. This temperature range allows the consistent retrieval of pyrolyzed products. The

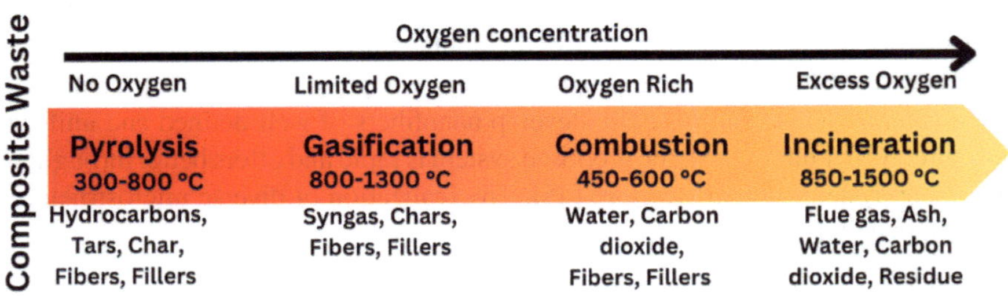

Fig. 7 A comparison of different thermal approaches for recycling of composite for chemical and energy recovery

process of pyrolysis has been carried out on various equipment depending upon the technology used such as conveyer pyrolysis, fluidized bed pyrolysis, catalytical pyrolysis, superheated steam pyrolysis, and microwave-assisted pyrolysis [46]. Conventional pyrolysis techniques demand energy of about 20–30 MJ/kg, while the energy demand of modern approaches is 5–10 MJ/kg. In classical pyrolysis, the composite waste is transported through a chain into the furnace at a controlled temperature and atmosphere as shown in Fig. 8. Alternatively, rotary furnaces are also used for the pyrolysis process as presented in Fig. 11. Solid wastes such as fibers, fillers, and char are collected from the bottom end of the furnace, and pyrolysis gas is passed through a condenser. Pyrolysis gas is liquified into pyrolysis oil that is collected from the condenser and combustible gases are fed to the afterburner for the recovery of heat to operate the boiler for steam generation. The temperature of the afterburner reaches 1200 °C. Fluidized bed pyrolysis is emerging as a promising technology owing to its ability to avoid the generation of toxic compounds [47]. Fluidized bed pyrolysis uses the same equipment as presented in Fig. 10. A fluidized bed allows the rapid heating of composite waste, as it involves the mixing of silica sand fluidized with hot air. There are several types of bed possibilities such as stationary, bubbling, vibratory, and circulating, and similarly there exists boiling, slugging, spouting, and channeling flow regimes. As-obtained hydrocarbons through pyrolysis would easily become the feedstock for the generation of monomers again through chemical processing. The ratio of the output is highly dependent on the type of composite waste and processing conditions. However, the average proportion of output involves 50–65% solid, 10–50% liquid, and 5–15% gases. The calorific value of gas is also linked to resin that is being pyrolyzed. The composition of pyrolysis oil is near to heavy crude or fuel oil. The calorific value of pyrolysis oil is around 37 MJ/kg and is the mixture of aromatic compounds and oxygenated. In the whole composition, 2/3 of the mixture is aromatic and 1/4 of the mixture is oxygen bearing organic compounds (carboxylic acid, and ketones). Solid pyrolysis residue is used as a filler after grounding through the sheet molding process. Although, they might not be as cheap as virgin fillers but being a byproduct, it is worth to use as a filler along with recovery of organic feedstock and heat. Pyrolysis offers a promising route for both recycling and reuse. The pyrolysis already attained a technology readiness level of 7–8. The rising crude oil prices are providing a chance to build up pyrolysis plants at an industrial scale.

8.2 Gasification

Gasification of composite waste falls in between pyrolysis and combustion as depicted in Fig. 7. In the process, grounded composite scrap is fed into the cracker in a limited oxygen supply at a temperature range of 1000–1300 °C. This transforms the scrap into small hydrocarbons that are further fed into a gasifier with steam to produce synthetic gas (syngas) as exhibited in Fig. 9. The outputs of the gasification are short hydrocarbon oil and gas, and syngas. The process of gasification transforms

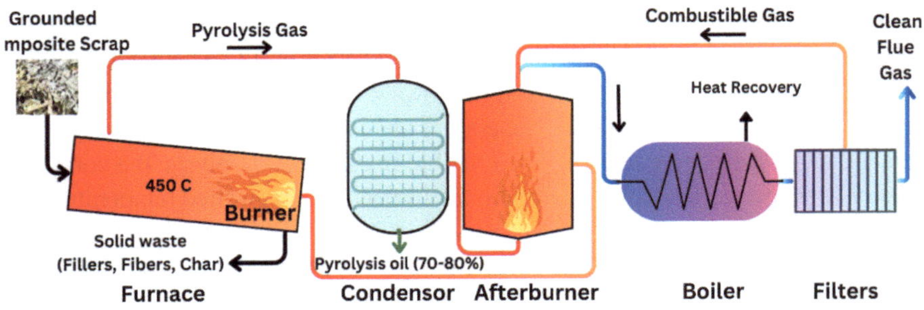

Fig. 8 Schematic of pyrolysis of composite waste

the SFRC completely into the hydrocarbons. It is a mature process and can be used in combination with pyrolysis.

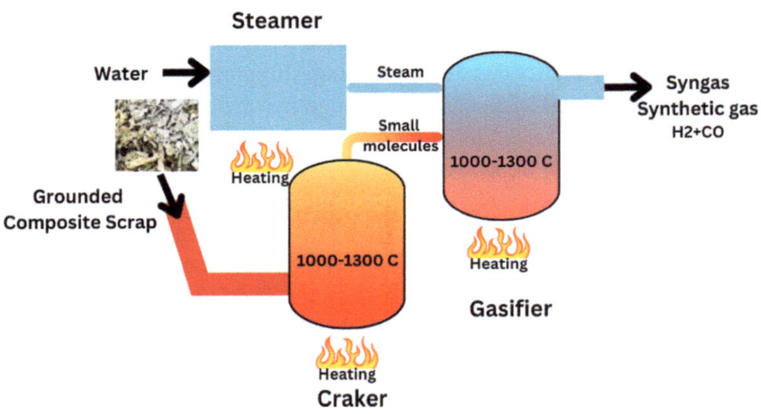

Fig. 9 Schematic of gasification of composite waste

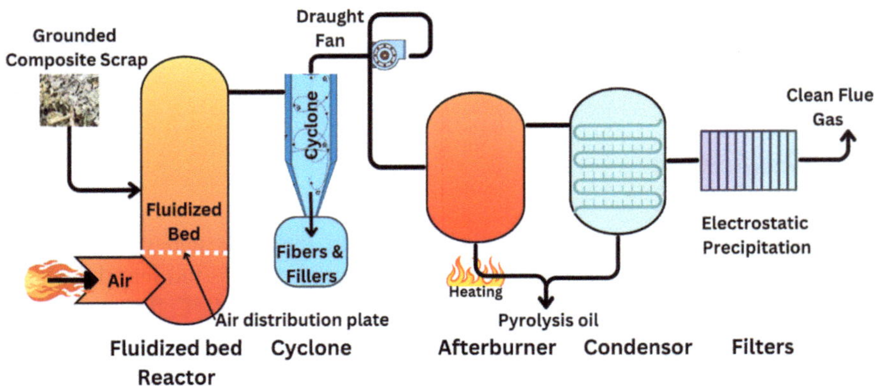

Fig. 10 Schematic of fluidized bed combustion and pyrolysis processes

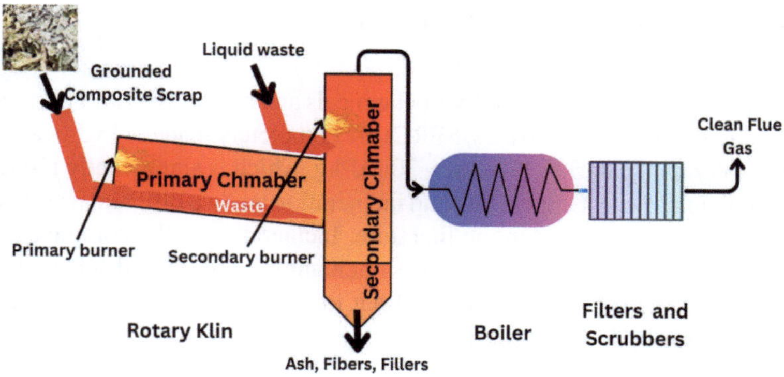

Fig. 11 Schematic of rotary kiln incineration process

9 Recycling of Blended/Hybrid SFRC

Synthetic fibers are also used in combination with other fibers such as basalt, carbon, or glass fibers to integrate the multifaceted properties into the composite products. Such hybrid composite structures are realized by layering the plies of different fibers in various configurations. Each layer of the composite imparts its characteristic property to the final composite [48]. The thermoset carbon fiber composite laminates are assembled using waste nylon bags as interlayers with varying thickness [49]. This approach of using nylon interlayers allows to realize 100% recyclable composites through thermos-forming. This approach also brings better toughness and enhanced damping properties to the composites. In a study, carbon and aramid fibers hybrid composites are fabricated using the cross-ply method for the construction of the helmet [50]. As-manufactured helmets exhibited better performance in terms of strength, heat resistance, and impact energy absorption than that of acrylonitrile butadiene (ABS) plastic. The recycling of such composites is even more challenging as compared to single fiber composites. Liu et al. have devised a novel sol–gel strategy to fabricate hybrid aramid nanofiber and wood fibers composites that are strong and tough owing to dynamic noncovalent cross-linking. These composites can be easily repaired under heating and recycled through dissolution and reconstruction without altering the ultimate mechanical properties [51]. Manoharan et al. have reported the integration of recycled aramid and basalt fibers into hybrid brake friction composite [52]. The inclusion of fibers in combination has improved the mechanical and thermal properties of the friction brakes [53].

10 Energy Recovery and Disposal

SFRC waste is essentially made of petrochemicals and can be used for heat recovery. Unlike inorganic fibers composites, SFRCs are completely made of hydrocarbon and provide a very high calorific value of > 30 MJ/kg, which is significantly higher than that of municipal waste. Combustion and incineration are two different processes and have been alternatively used in the literature. Incineration is the complete burning of waste at higher temperatures in excess amount of oxygen and left with ashes. Combustion on the other hand is carried out under mild conditions that allow the recovery of fibers and fillers too.

10.1 Combustion

The process of combustion is performed in a controlled environment and is carried out in a fluidized bed reactor as depicted in Fig. 10. A fluidized bed reactor is found suitable for the burning of several thermoset resins. This process allows energy recovery and recycling of fibers and fillers. The process temperature of combustion is reached to 450–600 °C which does not allow the damaging of the fibers [54]. The combustion process allows the recovery of glass and carbon fibers but is ineffective for SFRC owing to the organic nature of reinforced fibers. However, the combustion of SFRC allows the reclamation of inorganic fillers and metal insets that may be used in the construction of composite such as tires. In a typical reactor, grounded scrap is fed into the fluidized bed reactor with oxygen-rich air to start the burning. The burned residue is transported into a cyclone chamber that allows the separation of fibers and fillers. The unburnt organic compounds are then fed into an afterburner chamber that allows further recovery of heat. On passing through a condenser, pyrolysis oil is collected, and the flue gas is cleaned before sending to the environment by passing through an electrostatic precipitator. The flue gas also got a passage of scrubber to capture the SO_x and NO_x gases. The other advantage of combustion is that it does not require the prior removal of metal inserts owing to the attainment of low temperature of the process.

10.2 Incineration and Co-Incineration

Incineration and co-incineration are the complete burning of composite wastes in which the temperature of the process goes beyond 1000 °C. SFRC waste is essentially made of petrochemicals, hence, it can replace a significant fraction of conventional fossil fuels such as coal in the processes involving the use of kilns. In both processes, the energy is recovered equivalent to the calorific value of resin and synthetic fibers which is approximately equal to 30 MJ/kg. Incineration is either carried out in a grate-firing incinerator or a rotary kiln incinerator, while the co-incineration is conducted in a rotary cement kiln. The choice of incinerator is highly dependent on the type, volume, and hazard of the composite scrap. For example, a grate-firing incinerator is commonly used for municipal waste owing to its large volume availability. The grate combustion temperature reached 850 °C and even higher to 1100 °C using an auxiliary burner for hazardous waste. Solid discharge is collected at the bottom of the incinerator and flue gases are passed through a secondary combustion chamber to burn out volatile organic compounds (VOCs). Flue gases are passed through scrubbers and filters to clean it out to the environment. The incineration of composite scraps is performed with municipal waste at the plant site with heat and power recovery equipment [55]. The heat of the process is used to generate steam that is used to run a power plant for electricity generation and to feed the district heating system.

On the other side, a rotary kiln is relatively a robust system that allows the burning of any kind of waste as presented in Fig. 11. However, this system is mostly used for hazardous and industrial wastes [56]. The residence time of 30 to 90 min and the process temperature of 900–1200 °C is provided in the kiln for the complete bringing of the waste. The rotation of the kiln allows good burnout and prevents the generation of fine dust and particles. The kiln is further kept under a minor negative pressure that prohibits the escape of unburnt waste and the leakage of gases. Secondary (auxiliary) burning of flue gas is carried out in the secondary chamber to realize the burning of VOCs. The heat is used to generate steam through the boiler and flue gas is cleaned by passing through scrubbers and filters. SFRC does not cause any fiber residual problems as linked with glass, ceramic, and other inorganic fiber composites. The process of burning these wastes must meet strict specifications to limit the possible emissions of toxic gases and substances. For a very similar reason, such waste is also pre-processed to realize tailor-made fuels such as refuse-derived fuel (RDF) or solid recovered fuels (SRF). RDF pellets are produced from various types of waste including municipal solid waste, industrial waste, or commercial waste. RDF is obtained from mixed waste of which the major portion is combustible including non-recyclable plastic, labels, paper cardboard, etc. The process of producing RDF pellets starts with the shredding of waste followed by separation through manual sorting, size screening, magnetic, eddy current, infrared, and density separations. Then drying, mixing, homogenization, and pelletizing are performed to obtain RDF pellets. RDF pellets are fed along with coal in power plants or co-incinerated in cement kilns [57]. The ashes from incinerators can be either used in aggregates for the construction industry or land filled. The incineration of SFRC is quite advantageous as it produces minimal ash waste and does not produce fibers' fluff as is the case with inorganic fiber composites. The land-filling of ash waste brings about costs of land-filling and fibers' fluff is harmful to the workers' safety and health. Incineration is an industrialized process and has been used in several countries for the recovery of heat and power [58]. However, the challenges and drawbacks linked to inorganic fiber composites make it essential to collect those wastes separately and perform the co-processing.

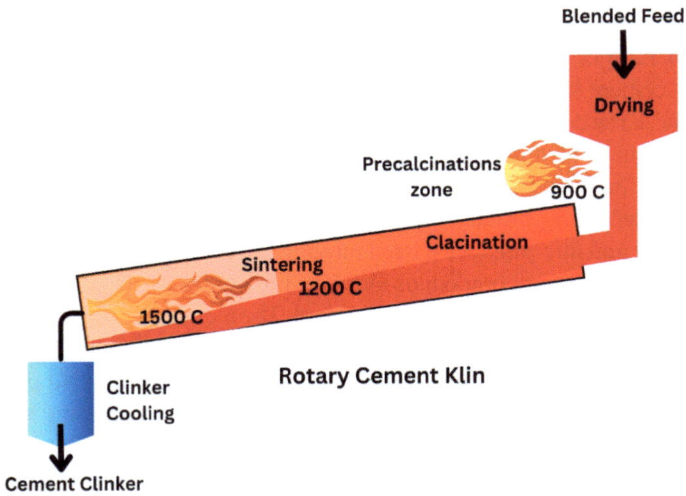

Fig. 12 Schematic of co-incineration (co-processing) of composite waste in a cement kiln

Coprocessing or co-incineration of composites in cement kiln is found to be an interesting solution to make the residual ashes as feedstock of cement production preserving the virgin resources [59]. The typical co-incineration processing of composite in a rotary cement kiln is depicted in Fig. 12. Pre-calcination allows the moisture removal, preheating, and decarbonization. The composites are burned to reach the temperature of 1500–2000 °C in a cement kiln. The inorganic residue coming from fibers, that is very much high for inorganic fibers-based composites, acts as an alternative source of raw material for cement production. The calcination and sintering of the fed raw materials give rise to clinker that is further grounded to cement. Hence, co-incineration prevents the requirement of landfilling and incorporates some value into the inorganic residue. This process has already been demonstrated at an industrial scale for its technical and commercial feasibility. However, this process is not an alternative to recycling but a good solution to get rid of composite waste by utilizing both organic and inorganic components of highly mixed scrap.

10.3 Landfill

Presently, most of the composite waste ends up in landfills, however, sustainability awareness and legislation such as extended producer responsibility and end-of-life directives will change the scenario in the next couple of years. European countries have already banned landfilling of plastic waste as it takes hundreds of years to degrade and become an integral part of the soil. Landfill taxation pushes the producer to work on the development of recycling technologies that have been discussed in detail. The landfill is only acceptable if there is a need to dispose of the composite waste that is not recyclable. However, the landfill of waste incurs a cost after the current legislation. In some countries, the landfill costs are more than the cost of mechanical recycling. However, landfill is still practiced in most countries.

11 Conclusion and Future Outlook

The recycling of SFRC is a critical area of research and industrial development, driven by the increasing demand for sustainable materials and the urgent need to manage composite waste. This chapter has explored the various types of SFRC, their applications, and the diverse recycling strategies available, including mechanical, chemical, dissolution, and thermal methods. The findings highlight that while progress has been made in recycling SFRC, significant challenges remain in achieving a fully circular economy for these materials. One of the key takeaways is that different recycling methods offer distinct advantages and limitations. Mechanical recycling is cost-effective and widely implemented but often leads to fiber degradation, reducing the mechanical performance of recycled composites. Dissolution and chemical recycling, including hydrolysis and solvolysis, provide more efficient fiber recovery but require the use of specialized solvents and catalysts, which may raise environmental and economic concerns. Thermal recycling, particularly pyrolysis and gasification, offers a means of energy recovery but does not contribute to material circularity in the same way as direct fiber reuse. Additionally, hybrid and blended SFRCs pose further challenges due to the complexity of separating different fiber types and resin matrices.

Moving forward, several key areas must be addressed to improve the sustainability of SFRC recycling. First, the development of new resin systems, such as dynamic covalent adaptable networks and vitrimer-based composites, could enable easier recycling without compromising mechanical properties. Second, advancements in sorting and processing technologies, including high-voltage fragmentation and solvent-based separation, could enhance the efficiency of fiber recovery. Third, integrating SFRC waste into industrial applications such as construction materials, automotive parts, and protective gear offers a practical route to increasing the market value of recycled fibers. In the broader context, policy interventions, extended producer responsibility (EPR), and stricter landfill regulations will play a crucial role in promoting SFRC recycling. Industries must adopt eco-design principles that prioritize recyclability at the product development stage. Furthermore, collaborations between academia, industry, and policymakers will be essential to bridge the gap between research innovations and large-scale implementation. Ultimately, the transition toward a more sustainable SFRC recycling framework will require a combination of technological innovation, regulatory support, and industry-wide commitment. By addressing current challenges and leveraging emerging solutions, the recycling of SFRC can contribute significantly to reducing environmental impact and fostering a circular economy in fiber-reinforced composites.

References

1. Unterweger C, Brüggemann O, Fürst C (2014) Synthetic fibers and thermoplastic short-fiber-reinforced polymers: properties and characterization. 35(2):227–236
2. Sundaramoorthy E, Mahanwar PA, Patil J, Mundhe G (2023) Polyolefin fiber, polyolefin fiber reinforced composites and their applications: a review. 43(3):219–230
3. Li W, Feng M, Liu X, Huang M, Ma R (2019) Ultra-high molecular weight polyethylene fibers/epoxy composites: effect of fiber treatment on properties. Fibers Polym 20(2):421–427
4. Senapati AK, Kutty SKN, Pradhan B, Nando GB (1989) Short polyester fiber reinforced natural rubber composites. Int J Polym Mater Polym Biomater 12(3):203–224
5. Meng X, Kang L, Guo X, Tang X, Liu L, Shen M (2023) Highly strong interface adhesion of polyester fiber rubber composite via fiber surface modification by meta-cresol/formaldehyde latex dipping emulsion. 15(4):1009
6. Gore PM, Kandasubramanian B (2018) Functionalized aramid fibers and composites for protective applications: a review. Ind Eng Chem Res 57(49):16537–16563
7. Zhou D, Huang H, Wang Y, Yu J, Hu Z (2020) Design and synthesis of an amide-containing crosslinked network based on Diels-Alder chemistry for fully recyclable aramid fabric reinforced composites. Compos Sci Technol 197:108280
8. Lin TR, Lin TA, Lin M-C, Lin Y-Y, Lou C-W, Lin J-H (2020) Using recycled high-strength polyester and Kevlar® wastes to reinforce sandwich-structured nonwoven fabric: structural effect and property evaluation. J Clean Prod 267:121899
9. Poulakis JG, Arvanitopoulos CD, Papaspyrides CD (1995) A technique for coating aramid fibers by recycling of aramid-based thermoplastic composites. 8(4):410–419
10. Vaidya U, et al (2024) Effect of recycled fibers and shredded intermediates variation on the mechanical properties and energy absorption of fiber-reinforced composite panels. 45(4):3700–3712
11. Prata JC, Silva ALP, Walker TR, Duarte AC, Rocha-Santos T (2020) COVID-19 pandemic repercussions on the use and management of plastics. Environ Sci Technol 54(13):7760–7765
12. Kwak JI, An Y-J (2021) Post COVID-19 pandemic: biofragmentation and soil ecotoxicological effects of microplastics derived from face masks. J Hazard Mater 416:126169
13. Signorini C, Marinelli S, Volpini V, Nobili A, Radi E, Rimini B (2022) Performance of concrete reinforced with synthetic fibres obtained from recycling end-of-life sport pitches. J Build Eng 53:104522
14. Bertelsen IMG, Ottosen LM (2022) Recycling of waste polyethylene fishing nets as fibre reinforcement in gypsum-based materials. Fibers Polym 23(1):164–174
15. Garg R, Garg R (2021) Performance evaluation of polypropylene fiber waste reinforced concrete in presence of silica fume. Mater Today Proc 43:809–816
16. Mohammadhosseini H et al (2020) Durability and thermal properties of prepacked aggregate concrete reinforced with waste polypropylene fibers. J Build Eng 32:101723
17. Evstatiev M, Fakirov S, Krasteva B, Friedrich K, Covas JA, Cunha AM (2002) Recycling of poly(ethylene terephthalate) as polymer-polymer composites. 42(4):826–835
18. Su J, Jiang Z, Fang C, Yang M, Wu L, Huang Z (2022) The reinforcing effect of waste polyester fiber on recycled polyethylene. 14(15):3109
19. Miao X et al (2024) Research on the application of a PET/PP composite board prepared by recycling waste polyester clothing in furniture products. ACS Sustain Resource Manag 1(5):890–897
20. Weng L, Zhang X (2023) Recycling polyester fiber with bio-based alginate fiber into fire-safety composite. Polym Degrad Stab 217:110519
21. Moghaddamzadeh S, Rodrigue D (2018) The effect of polyester recycled tire fibers mixed with ground tire rubber on polyethylene composites. Part I: Morphol Anal 34(4):200–220
22. Moghaddamzadeh S, Rodrigue D (2018) The effect of polyester recycled tire fibers mixed with ground tire rubber on polyethylene composites. Part II: Physico-Mech Anal 34(3):128–142
23. Kazemi H, Fazli A, Ira JP, Rodrigue D (2023) Recycled tire fibers used as reinforcement for recycled polyethylene composites. 11(9):74

24. Bocci E, Prosperi E (2020) Recycling of reclaimed fibers from end-of-life tires in hot mix asphalt. J Traf Transport Eng (English Edition) 7(5):678–687
25. Teijin Frontier Co. L (2022) Teijin Frontier develops world's first tire cord made from eco-friendly adhesive and recycled polyester. https://www2.teijin-frontier.com/english/news/post/128/
26. Dorigato A, Fambri L (2013) Effect of aramid regenerated fibers on thermo-mechanical behaviour of polyamide 12 composites. 32(17):1243–1256
27. Kouparitsas CE, Kartalis CN, Varelidis PC, Tsenoglou CJ, Papaspyrides CD (2002) Recycling of the fibrous fraction of reinforced thermoset composites. 23(4):682–689
28. Loureiro L, Henrique Carvalho V, Prado Bettini SH (2016) Reuse of p-aramid from industrial waste as reinforcement fiber in polyamide 6.6. Polym Test 56:24–130
29. Chuang Y-C, Bao L, Lou C-W, Lin J-H (2019) Hybrid-fiber-reinforced composite boards made of recycled aramid fibers: preparation and puncture properties. Fibers Polym 20(2):398–405
30. Ma S, Li H, Li C, Li B, Fei J, Wen Y (2022) Valuable aramid/cellulose nanofibers derived from recycled resources for reinforcing carbon fiber/phenolic composites. Carbohyd Polym 292:119712
31. Garfias K, Odnevall I, Odelius K, Hakkarainen M (2023) Surface modification of aramid fiber meshes—the key to chemically recyclable epoxy composites. RSC Sustain 1(8): 1967-1981. https://doi.org/10.1039/D3SU00258F
32. Alabiso W, Schlögl S (2020) The impact of vitrimers on the industry of the future: chemistry properties and sustainable forward-looking applications. Polymers 12(8):1660
33. Maschino J, Rougnon-Glasson O (2015) Procede de recyclage de pieces en composite thermoplastique. Patent FR3007684A1
34. Fazli A, Rodrigue D (2020) Recycling waste tires into ground tire rubber (GTR)/rubber compounds: a review. 4(3):103
35. Mativenga PT, Shuaib NA, Howarth J, Pestalozzi F, Woidasky J (2016) High voltage fragmentation and mechanical recycling of glass fibre thermoset composite. CIRP Ann 65(1):45–48
36. Anthony WS (2006) Technology to separate rubber crumb from fiber. Appl Eng Agric 22(4):563–570
37. Hashimoto A, Satoh M, Iwasaki T, Morita M (2002) Fibrillation of aramid fiber using a vibrating ball mill and evaluation of the degree of fibrillation. J Mater Sci 37(18):4013–4017
38. Garfias K, Hakkarainen M, Odelius K (2024) Mechanical recycling of epoxy composites reinforced with short-cut aramid fibers: surface functionalization—the missing piece of the puzzle. Polymer 295:126747
39. Liao Y, Chen C, Wei B, Yang S, Bai S (2022) Green recycling of aramid fiber waste as the friction modifier of POM by solid state shear milling technology. 33(10):3540–3550
40. An W et al (2023) Water-solvent regulation on complete hydrolysis of thermosetting polyester and complete separation of degradation products. J Hazard Mater 453:131423
41. Wu Z et al (2024) Efficient volumetric hydrolytic degradation of epoxy resin driven by permeability-enhanced surfactants. Chem Eng J 498:155528
42. Yuyan L, Li L, Linghui M (2009) The experimental research on recycling of aramid fibers by solvent method. 28(18):2211–2220
43. Kirshanov K, Toms R, Melnikov P, Gervald A (2022) Investigation of polyester tire cord glycolysis accompanied by rubber crumb devulcanization. Polymers (Basel) 14:4
44. Yang B, Li W, Zhang M, Wang L, Ding X (2021) Recycling of high-value-added aramid nanofibers from waste aramid resources via a feasible and cost-effective approach. ACS Nano 15(4):7195–7207
45. Chen H-J, Bai Q-Y, Liu M-C, Wu G, Wang Y-Z (2021) Ultrafast, cost-effective and scaled-up recycling of aramid products into aramid nanofibers: mechanism, upcycling, closed-loop recycling. Green Chem 23(19): 7646-7658. https://doi.org/10.1039/D1GC01805A
46. Naqvi SR, Prabhakara HM, Bramer EA, Dierkes W, Akkerman R, Brem G (2018) A critical review on recycling of end-of-life carbon fibre/glass fibre reinforced composites waste using pyrolysis towards a circular economy. Resour Conserv Recycl 136:118–129

47. Kaminsky W (2010) 8—Fluidized bed pyrolysis of waste polymer composites for oil and gas recovery. In: Goodship V (ed) Management, recycling and reuse of waste composites. Woodhead Publishing, pp 192–213
48. Raci Aydin M, Acar V, Cakir F, Gündoğdu Ö, Akbulut H (2022) Comparative dynamic analysis of carbon, aramid and glass fiber reinforced interply and intraply hybrid composites. Compos Struct 291:115595
49. Proietti A, Quadrini F, Gallo N, Santo L (2023) Thermo-formable hybrid carbon fibre laminates by composite recycling. Int J Adva Manuf Technol 127(5):2989–3001
50. Kim S, Lee J, Roh C, Eun J, Kang C (2019) Evaluation of carbon fiber and p-aramid composite for industrial helmet using simple cross-ply for protecting human heads. Mech Mater 139:103203
51. Liu T et al (2023) A novel sol-gel strategy for constructing wood fibers and aramid nanofiber nanocomposite with strong, tough and recyclable properties. Compos Sci Technol 238:110026
52. Manoharan S, Vijay R, Singaravelu DL, Krishnaraj S, Suresha B (2019) Tribological characterization of recycled basalt-aramid fiber reinforced hybrid friction composites using grey-based Taguchi approach. Mater Res Express 6(6):065301
53. Manoharan S, Shihab AI, Alemdar ASA, Babu LG, Vijay R, Lenin Singaravelu D (2019) Influence of recycled basalt-aramid fibres integration on the mechanical and thermal properties of brake friction composites. Mater Res Express 6(11):115310
54. Yang Y, Boom R, Irion B, van Heerden D-J, Kuiper P, de Wit H (2012) Recycling of composite materials. Chem Eng Process 51:53–68
55. Job S, Leeke GA, Mativenga PT, Oliveux G, Pickering SJ, Shuaib NA (2016) Composites recycling—where are we now?
56. Williams PT (2005) Other waste treatment technologies: pyrolysis, gasification, combined pyrolysis–gasification, composting, anaerobic digestion. In: Waste treatment and disposal. pp 325–366
57. Velis CA, Cooper J (2013) Are solid recovered fuels resource-efficient? Waste Manage Res 31(2):113–114
58. Rybicka J, Tiwari A, Leeke GA (2016) Technology readiness level assessment of composites recycling technologies. J Clean Prod 112:1001–1012
59. Vijay N, Rajkumara V, Bhattacharjee P (2016) Assessment of composite waste disposal in aerospace industries. Procedia Environ Sci 35:563–570

Recycled Blended/Hybrid Fiber-Reinforced Composites (RBRFC)

Aima Sameen Anjum

1 Introduction

Recycled Fiber-Reinforced Composites (RBRFCs) are made using fibers from waste or post-consumer products, combined with virgin or other recycled fibers through various recycling methods. These composites include recycled glass fibers from wind turbine blades and construction waste, recycled carbon fibers from aerospace and automotive parts, and natural fibers from agricultural waste. They offer environmental sustainability by reducing raw material needs and landfill waste, cost-effectiveness, and resource efficiency. Applications span automotive, construction, aerospace, and consumer goods, though challenges like quality control and scaling production remain. Future advancements in technology and recycling processes aim to enhance their performance and integration [1].

2 Composite Fabrication Techniques

Composite fabrication is done in several ways nowadays. Still, their choice depends on the useful facilities, composite kind, applied materials, financial budget rate, and composite usage. Basically, every manufacturing technique has benefits and drawbacks. For instance, some techniques are time-consuming and have low composite manufacturing prices while others go swiftly and have great expenses. Furthermore, intimately linked with the production cost are the precision and quality of the produced composite. It is interesting to note that not all fabrication techniques fit every kind of composite as some need specific molds to shape the reinforcing

A. S. Anjum (✉)
National Textile University, Faisalabad, Pakistan
e-mail: aimasameen@ntu.edu.pk

H. Jamshaid and U. Zubair (eds.), *Sustainable Recycling of Fiber Reinforced Polymer Composites*, SDGs and Textiles, https://doi.org/10.1007/978-981-96-9536-2_7

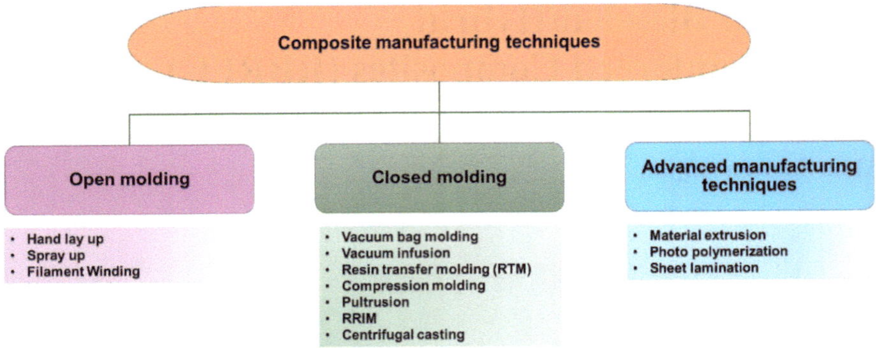

Fig. 1 Composite fabrication methods

and matrix phases into the intended final form. Composite fabrication techniques are listed as illustrated in Fig. 1.

2.1 *Open Molding*

Hand lay-up provides one of the most straightforward methods for creating composites using the thermoset matrix phase. During the production process, the fiber plies are manually positioned next to one other until the laminate forms; resin is then used to ensure their adherence. After the fiber plies are first filled with resin, they are bundled using a different method called wet hand lay-up.

Hand lay-up and spray-up techniques are almost identical. But in this technique, the resin and chopped fibers are sprayed through the spray gun on a mold. These fibers are fused into the matrix material at the same time using a roller. The procedure is shown in Fig. 2. Chopped fibers offer high conformability in an open mold approach, which is also significantly quicker than hand lay-up. Nevertheless, this above construction technique has lost acceptability due to the styrene (a harmful pollutant) content in the resin.

It is a computerized open molding technique in which a rotating mandrel is used as mold to manufacture hollow tube and cylindrical components, i.e., golf shafts, car shaft drivers, and spaceship parts construction. As shown in Fig. 2, the reinforced fibers that were taken out of the roving are sent into the hot resin bath. To create the required product shape, the fibers soaked with resin are wrapped around a revolving mandrel. After applying enough layers, the mandrel is taken off and the laminate is ready to cure. The fundamental benefit of filament winding—which has drawn a lot of interest—is cheap manufacturing cost. This process is advantageous in providing excellent tensile strength, uniformity in products, and strength to weight ratio.

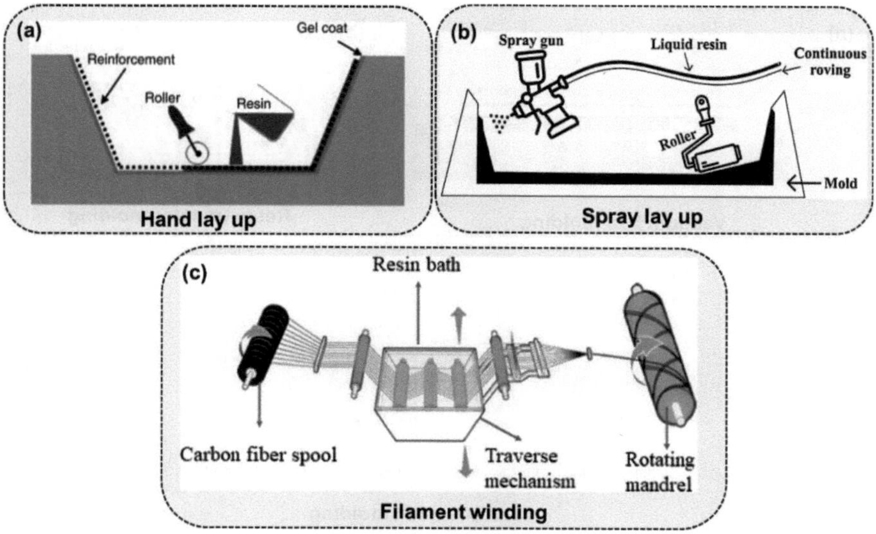

Fig. 2 Open mold fabrication methods, **a** hand lay-up, **b** spray lay-up, **c** filament winding

2.2 Closed Molding

In vacuum bag molding, the reinforcement is infused with resin and a thin layer of synthetic fibers (nylon, PVA, etc.) is applied over wet lay-up. After that the edges are sealed and the vacuum bag is mounted on the mold as presented in Fig. 3. Higher levels of fiber reinforcement are then achieved by forcing the hoover bag to consolidate the laminate by using external atmospheric pressure to remove the trapped air and extra resin. A hoover technique was used to consolidate a large number of inorganic laminate composites that were created by hand lay-up.

Vacuum infusion and vacuum bag molding are very similar techniques with slight difference as vacuum is established prior to the injection of resin. Another name for it is the resin infusion method. After a vacuum bag is positioned and sealed at the edge of the mold, a perforated tube is positioned between it and the resin container. The dry fiber reinforcement is then covered with a porous membrane, and the laminates are consolidated by applying the pressure of vacuum to suck in the resin. Laminated shrinks when resin is pulled into the mold, preventing extra resin from accumulating.

Resin transfer molding method has grown in popularity because of its effectiveness, affordability, and little environmental impact. Since RTM uses dry textiles and raw resin directly at room temperature, it provides immediate cost reductions throughout production when compared to alternative manufacturing techniques like autoclave curing. In this procedure, dry fiber reinforcing components are pre-placed in a mold cavity, and low viscose resin is injected under pressure. Because RTM is flexible, several fiber orientations may be used including 3D reinforcements, giving

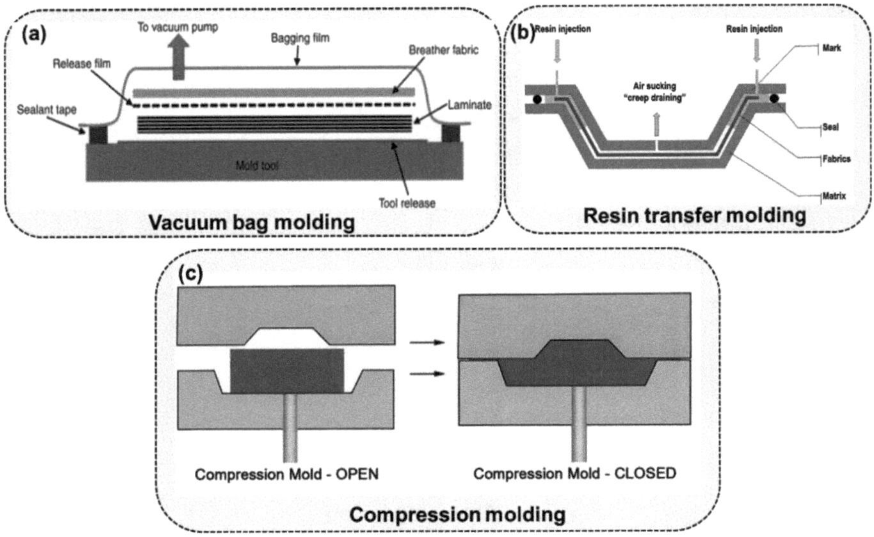

Fig. 3 Closed mold fabrication methods, **a** vacuum bag molding, **b** resin transfer molding (RTM), **c** compression molding

designers the freedom to customize the final composite product's characteristics as presented in Fig. 3.

Compression molding works best for complex fiberglass-reinforced polymer products because it is a high-volume, high-pressure process with a short cycle time as presented in Fig. 3. It makes use of metal molds that are placed on a mechanical or hydraulic molding press and heated to between 250° and 400 °F. The structure is made by first holding the resin charge in a mold, with one part of the mold fixed and the other section of the mold that is movable hot-pressed. There are several varieties of compression molding, including sheet molding compound (SMC), bulk molding compound (BMC), thick molding compound (TMC), and wet lay-up compression molding, depending on the kind of material that has to be molded. The difficulty of adding a significant amount of bamboo fibers to epoxy resin is overcome by using the hand lay-up method, which is followed by compression molding.

Pultrusion is a sustainable manufacturing technique in which FRP elements with a constant cross-section can be produced. It is a continuous process that can be tailored to both basic and complicated cross-sectional forms and has extremely good structural qualities. A tractor mechanism pulls continuous strands of glass, carbon, or basalt fiber soaked in a resin solution via a steel die. In order to solidify the saturated reinforcement and establish shape of stock, the steel die was heated quickly as shown in Fig. 4.

Reinforced reaction injection molding (RRIM) is very similar to RTM. Two or more resins are mixed together in a mixing vessel to produce a thermosetting polymer at high pressure. The combination is supplemented with reinforcement materials such as mica or glass fibers. The resin mixture is then metered into a mold using high

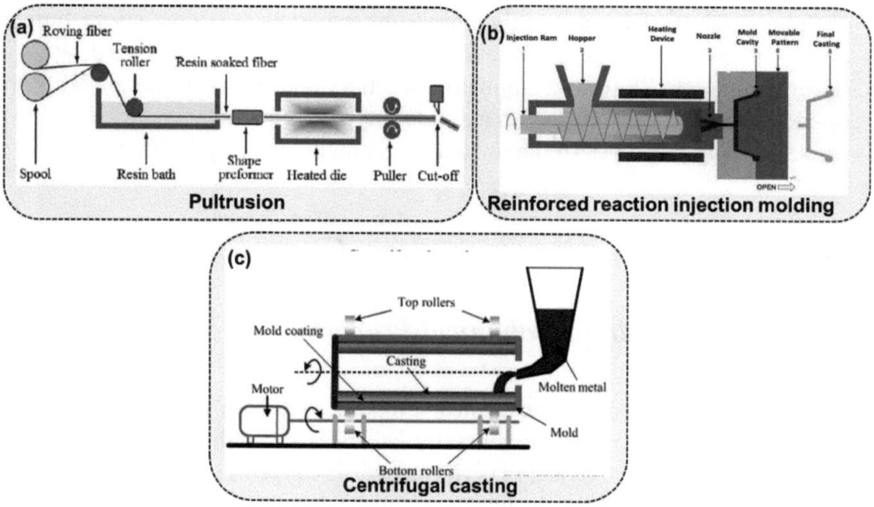

Fig. 4 Closed mold fabrication methods, **a** pultrusion, **b** reinforced reaction injection molding (RRIM), **c** centrifugal casting

pressure pumps or injection cylinders. The resulting liquid has a low viscosity even though it was combined under high pressure. The mixed resin with low viscosity is easily poured into the mold cavity at a comparatively low pressure of about 50 psi (Fig. 4). Within the mold cavity, polymerization happens very quickly without gaining any more heat.

In the centrifugal casting process, reinforcement and resin are placed inside the cylindrical rotating mold's surface. The mixture is held against the cylindrical mold's inner wall by centrifugal force until it cures. Figure 4 illustrates the process by pouring molten resin into a basin that fills the mold cavity. The centrifugal casting technique is used to produce nano size TiB2 particles inside an aluminum matrix while overcoming the low particulate volume % material restriction.

2.3 *Advanced Manufacturing Techniques*

The above-mentioned traditional manufacturing processes need molds which restrict the formability and make it expensive and bit difficult when there are complex structures. Whereas additive manufacturing (AM) employes a fabrication method in which composite structure is made layer by layer through computer aided designing process.

- Material extrusion: This method is also known as fused filament fabrication (FFF) as the fiber is melted and then extruded in this process. A nozzle is used to deposit the material in solid filament or in paste form.

- Vat photo polymerization: In this process the materials selection is very limited as it prints in high resolution. To cure a photopolymer, UV light is used.
- Sheet lamination: To form a laminated structure, thin layers of fabric sheets are bonded together. This is categorized in two types: (a) composite based additive manufacturing (CBAM), (b) laminated object manufacturing (LOM).

3 Properties Comparison with Virgin Composites

3.1 Factors Affecting Mechanical Properties

Several factors influence the mechanical properties of both recycled and virgin composites: Higher fiber content generally enhances strength, but excessive fiber can lead to agglomeration, reducing performance. The type of polymer matrix (thermoplastic or thermoset) affects the final mechanical properties, especially in recycled composites. Recycling processes, such as mechanical grinding or chemical dissolution, can lead to fiber degradation, impacting mechanical performance. Optimized processing can preserve fiber length and enhance properties.

3.2 Mechanical Properties

The right choice of composites for structural purposes is mostly determined by their mechanical characteristics, such as their impact resistance, flexural strength, and tensile strength. Depending on the quality of the recycled fibers and the recycling method employed, recycled composites may behave mechanically differently from virgin composites. Table 1 presents a comparison of the mechanical properties from various studies on recycled and virgin composites.

Tensile Strength: In general, the tensile strength of composites reinforced using recycled fibers is lower than that of their virgin alternatives. For example, recycled carbon fiber composites can maintain between 60 and 70% of their virgin carbon fiber composites' tensile strength. The obvious reason for reduction is the result of fiber degradation during recycling, which can cause defects, weak fiber-matrix interaction and reduction in fiber lengths a weaker matrix-fiber link. Composites made of virgin fibers usually have higher tensile strengths because the fibers are intact, preserving their integrity and giving the matrix superior reinforcement.

Flexural Strength: Because of the possibility of gaps or contaminants from the recycling process, as well as inferior fiber-matrix adhesion, recycled composites often have lower flexural strengths (resistance to bending). According to studies, the flexural strength of recycled glass fiber composites might be 10–30% less than that of virgin composites. On the other hand, because the fibers are precisely incorporated into the matrix, virgin composites exhibit constant and superior flexural qualities.

Table 1 Comparative analysis of mechanical properties in recycled and virgin composites using different recycling methods [5]

Authors	Recycling material	Recycling method	Recycled strength (MPa)	Virgin strength (MPa)
ATI	Carbon fiber reinforced composite	Chemical recycling	3325	3500
Hernanz et al.	Carbon fiber reinforced composite	Solvolysis	2975–3465	3500
Hyde et al.	Carbon fiber reinforced composite	Solvolysis	3900	4090
Liu et al.	Carbon epoxy composite	Solvolysis	3461	3500
Palmer	Fiber reinforced polyester sheet	Mechanical recycling	1130–2310	1490–2670
Takahashi et al.	Carbon fiber reinforced plastic	Thermal recycling	100	120
Kouparitsas et al.	Glass/ polypropylene thermoset composite	Mechanical recycling	25	24
Bernasconi et al.	Glass fiber composite	Mechanical recycling	185.3	187.4
Pickering et al.	Glass fiber thermoset composite	Fluidized-bed combustion	1750	3500
Kennerly et al.	Polyester sheet molding compound	Fluidized-bed combustion	1750	3500
Turner et al.	Scrap prepreg	Fluidized-bed combustion	1750–2625	3500
Pimenta and Pinho	Carbon fiber reinforced polymer	Pyrolysis	3200	4800

Impact Resistance: Impact resistance is yet another possible drawback of recycled composites. Shorter fiber lengths and specific surface flaws that make the material more brittle under abrupt pressures may reduce this characteristic. However, this problem might be resolved by employing specific recycling techniques that maintain the length and quality of the fiber [2].

3.3 Physical Properties

Basic characteristics like density, water absorption, and heat resistance have a significant impact on how well composites perform in a variety of scenarios.

Density: Variations in the density of composites can be produced via recycling procedures. Sometimes the existence of voids or shorter-length fibers cause recycled composites to have a minor decrease in density. Innovative methods, however, might reduce these density variations. Virgin composites typically maintain a consistent density throughout manufacturing, which is important for applications where weight and balance are critical.

Thermal Stability: Some thermal resilience may be lost by recycled composites, especially if the recycled fibers have been exposed to heat damage during production. For instance, compared to virgin composites, recycled glass fiber composites have somewhat less heat resistance. These composites frequently offer better heat stability since the fibers and matrix are less likely to degrade after initial production.

Water Absorption: Recycled composites may absorb more water if they have voids, microcracks, or a weaker fiber-matrix link. Over time, mechanical qualities may deteriorate due to increased water absorption. Virgin composites often retain less fluid as they have more fiber-matrix cohesion and fewer spaces in them [3].

3.4 Durability and Aging

For composites to be used practically, they must be long-lasting and resistant to environmental deterioration.

Longevity: The deterioration and poor bonding in recycled composite materials make them not last as long as virgin composites. However, durability has increased due to advancements in surface treatment and length of fiber retention caused by recycling processes. Virgin composites have a reputation for being strong and dependable, which enables them to endure throughout time.

Environmental Performance (UV, Moisture): Recycled composites are often more susceptible to environmental factors such as UV radiation and moisture. These conditions may speed up breakdown, especially if the recoverable fibers are less resilient due to the recycling process. Surface conditioning and protective coverings help mitigate this issue. Virgin composites' exceptional resistance to environmental influences makes them ideal for high-performance and outdoor applications.

3.5 *Cost-Efficiency and Economic Viability*

Cost Analysis: The cost of recycled fiber composites can be as much as 30–50% lower than that of virgin composites. The main reason for the lower cost is the reduced need for raw materials. The level of quality of the recycled product can differ and additional processing may be needed to match the performance of virgin composites. Virgin composites are suitable for challenging situations where performance must be compromised because they are more expensive but provide consistent quality.

Long-Term Benefits and Sustainability: Through waste reduction, resource conservation, and the promotion of circular economy principles, the usage of recycled composites advances sustainability objectives. The market for recycled composites is expected to grow as environmental restrictions become more stringent, which will spur the creation of more effective recycling technology. Despite their higher cost, virgin composites continue to dominate in applications where safety, performance, and reliability are paramount [4].

4 Recycling Techniques for RBRFC

Over the past decade, various studies have explored different methods for recovering fibers from thermoset composite materials for potential applications as shown in Fig. 5. Following is the detail of the methods used for recycling of composite materials.

4.1 *Mechanical Recycling*

Mechanical recycling methods have the potential to produce other raw materials through various stages of mechanical treatment and separation. This includes processes such as screening, straining, shredding (chopping), densification, milling, collection, and similar techniques as shown in Fig. 6. Cutting and crushing leftover or discarded composite parts into tiny pieces—often referred to as produced scraps—are the first steps in the mechanical recycling process. Materials for construction like artificial wood, asphalt, or cement mineral resources are made from powders and fibrous components that are separated from created scraps via straining. However, since these recycled products are cost-effective, methods to enhance the quality of recycled fibers (RFs) are necessary to maintain the quality of the developed products.

The mechanical recycling technique involves grinding materials into finer pieces after an initial crushing or shredding step. Different sizes of recyclates are separated by sieving into resin-rich powders and fibers of various lengths. Mechanical grinding is commonly applied to glass fiber reinforced composites (GFRP), particularly Sheet Molding Compound (SMC) and Bulk Molding Compound (BMC), but also to carbon

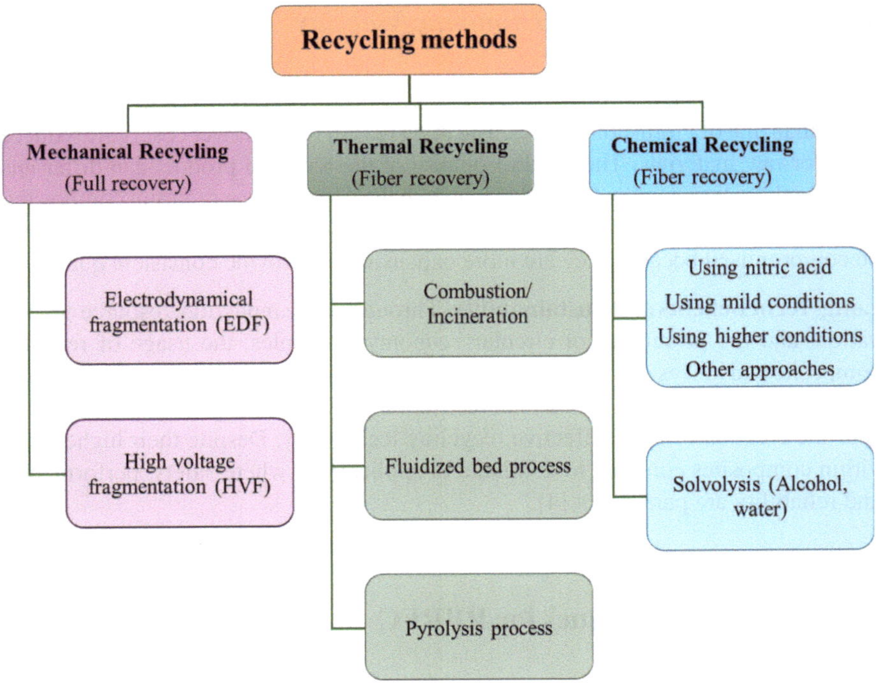

Fig. 5 Recycling methods

fiber reinforced composites (CFRC) [7, 8]. Although ground composite materials can be employed as reinforcement or fillers, their usage as fillers is not economically feasible because virgin fillers are so inexpensive. The incorporation of filler material is limited to less than 10% by weight to avoid deterioration in mechanical properties. Ground materials can also be used as an energy source due to their resin content. Studies have shown that cutting mills produce more homogeneous fiber lengths than hammer mills. Ground materials can be successfully incorporated into virgin resins, but the fibrous fractions of ground thermoset materials are challenging to reuse as reinforcement due to poor bonding with new resin. However, improved mechanical properties can be achieved with longer mixing times.

Manufacturers of glass fiber reinforced composite (GFRC), such Filon Products Ltd. in the UK and Mixt Composites Recyclables (M-C-R) in France, are the main industrial users of this method. Currently, there is no industrial grinding process for carbon fiber reinforced composites (CFRCs). Instead, other methods that separate fibers from the matrix are preferred for carbon fibers, as they allow for contamination-free recovery and potential reuse in new composites. Currently, a new method for shredding carbon fiber reinforced thermoplastic has been proposed: electrodynamic fragmentation. Using this technique, the material is submerged in water among two electrodes and is broken up into smaller pieces by applying a high voltage (50–200 kV) [6].

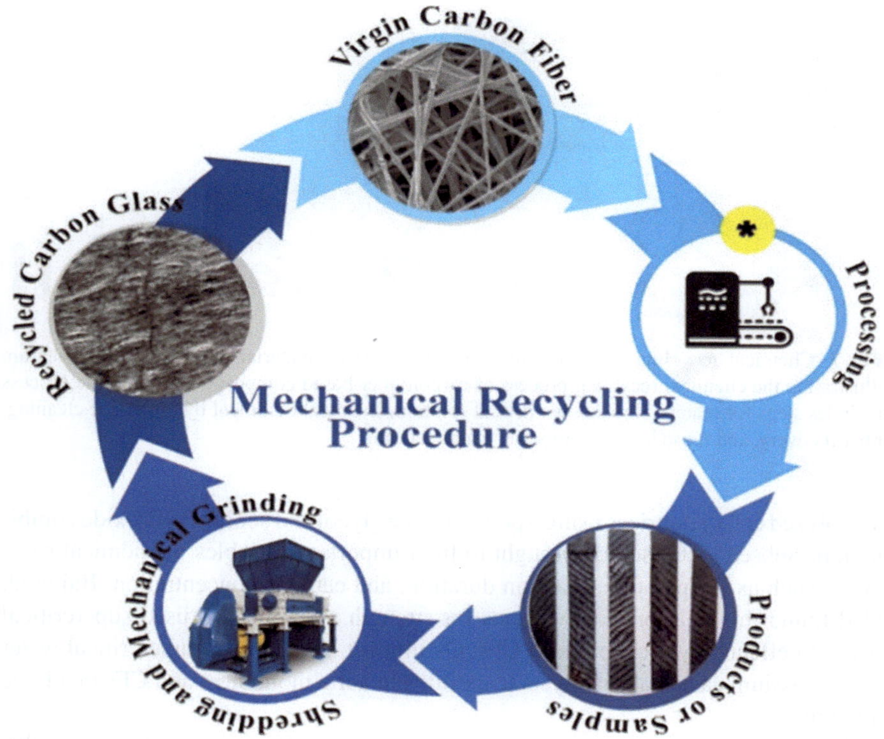

Fig. 6 Mechanical recycling process of the carbon fiber reinforced composite [6]. The schematic diagram illustrates the flow of the carbon fiber recycling process. Initially, virgin fiber is processed to create a composite. After its use, the material undergoes shredding or grinding to initiate recycling, ultimately resulting in recycled carbon fibers through a mechanical recycling process

4.2 *Chemical Recycling*

Fibers can be recovered with most of their mechanical and physical characteristics still present because chemical recycling converts polymer matrices into usable chemical solvents. Solvents used for decomposing epoxy resin matrices include tetralin, supercritical/subcritical fluids, nitric acid, and supercritical/subcritical water as shown in Fig. 7. Supercritical fluids, which have strong solvent properties, are particularly effective in this process. When above their critical point, these fluids enhance mass transfer rates due to their low viscosity and high diffusion constants. Bledzki et al. [9] studied the impact of various factors on fiber strength and damage in composites, finding that corrosive environments significantly reduce the fatigue resistance of E-glass fiber laminates. Torres et al. [10] investigated pyrolysis for recycling sheet molding compound (SMC), noting that temperatures above 400 °C did not significantly affect the outcome. Using subcritical and supercritical alcohols, Pinero-Hernanz et al. [11] demonstrated the carbon fibers' recycling with superior strength retention using sub-critical and supercritical alcohols. Yang et al. [12] effectively

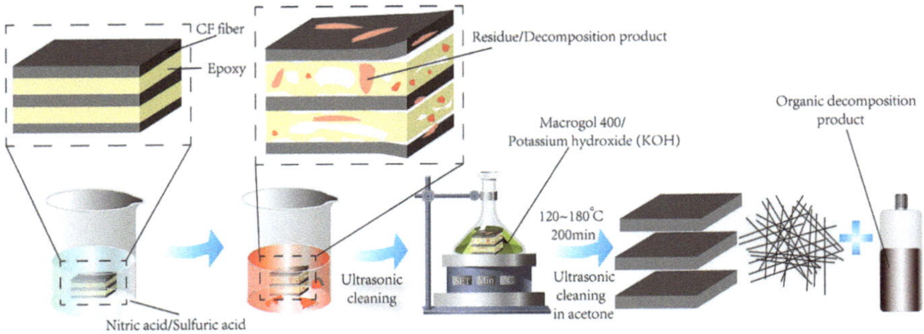

Fig. 7 Chemical recycling of carbon fiber based composite material [15]. Schematic diagram illustrating the chemical recycling process of carbon fiber-based composite material. The process includes steps for material breakdown in acid solution, chemical treatment by ultrasonic cleaning, fiber recovery, and reuse in new composite manufacturing

solvolyzed epoxy matrices using a polyethylene glycol and sodium hydroxide combination. Subsequent research brought to light important variables in chemical recycling, such as temperature, reaction duration, and catalyst concentration. Bai et al. [13] found the ideal breakdown rates for strength retention and used supercritical water to effectively recover clean CFs. Okajima et al. [14] used supercritical water and potassium carbonate catalysts to extract phenolic monomers and CFs in a large amount.

In chemical recycling operations, supercritical fluid water may dissolve bisphenol that has been hardened with diaminodiphenylmethane at a rate more than 95%. The use of potassium hydroxide and phenol can accelerate this pace even further. Water's structure is defined by its static dielectric constant, which drastically drops in the supercritical zone. This weakens hydrogen bonds and causes water to act more like a weakly polar or non-polar solvent. This characteristic makes it simple for water to combine with organic substances and turn them into a single phase. Unsaturated polyester resin matrices reinforced with long glass fibers (GFs) have been broken down by hydrolysis procedures employing subcritical water. The most important factor is temperature since mechanical qualities deteriorate more at higher temperatures. Due to the hazardous solvents and catalysts used, solvolysis—a chemical process that uses solvents to break down resin and recycle fibers—is an efficient but expensive and environmentally problematic procedure.

Clean CFs can be recovered by chemical recycling procedures that use water as a reaction medium at high temperatures. Epoxy resin in composites has been dissolved by hydrolysis with sub-critical water, producing pure GFs. High breakdown rates and clean fiber surfaces have been achieved in the recycling of CFs using a two-step method that includes an acetic acid pre-treatment and a DMF/H_2O_2 solution. With recovered CFs maintaining the majority of their initial tensile strength, the selective cleavage of C–N bonds in epoxy resin utilizing a breakdown system of $AlCl_3$ and CH_3 COOH has proven effective. Significant tensile strength retention in regenerated CFs has been achieved through the exploration of electrochemical techniques employing sodium hydroxide (NaOH) solutions. Furthermore, CFs have been recovered using supercritical 1-propanol; research has focused on how temperature, processing duration, and additives affect recycling effectiveness [16].

4.3 *Thermal Recycling*

For the purpose of recycling fiber-reinforced polymer composites (FRPCs), many thermal recycling methods have been developed. The fluidized bed process, combustion, and pyrolysis are the three primary thermal techniques. Waste composites are broken down into solid, liquid, and gaseous products by external heating in these procedures. Since most polymeric materials possess calorific value, the energy generated during these processes can be harnessed, making it possible to convert composite waste into electricity or other by-products. Pyrolysis includes the heat decomposition of composite waste, producing high energy owing to the polymeric elements present. The fluidized bed process effectively recovers fibers and other by-products from combined and polluted end-of-life (EOL) composites by curing the waste in hot air (450–550 °C) or in an oxygen-free environment (as in pyrolysis) [6].

Pyrolysis: Pyrolysis is a key thermal recycling technique for the degradation of organic components in composite waste, converting them into gases and liquids, while leaving behind solid residues such as fillers and fibers. Typically conducted in low or no oxygen environments, the process operates at temperatures ranging from 400 to 1000 °C, depending on the composite and reinforcement materials. The resulting by-products include tar and other heavy liquids that can serve as oils, and gases such as CO_2, H_2, and hydrocarbons, which can be utilized as energy resources due to their calorific value. Optimizing process parameters is necessary for efficient pyrolysis in order to guarantee the highest possible recovery of materials, fuels, and chemical products. Glass fibers (GFs) within thermoset composites can be recycled via pyrolysis, but the strength of the fibers is much reduced, which results in reduced mechanical performance in new composites. At temperatures ranging from 300 to 700 °C in nitrogen atmospheres, recovered GFs exhibited a 50% drop in tensile strength.

In order to minimize char deposition on recycled carbon fibers (RCFs), Meyer et al. [17] stressed the significance of regulating temperature, dwell time, and environmental factors (such as nitrogen vs. synthetic air). They discovered that partial oxidation at 500 °C for two hours might produce clean RCFs with characteristics similar to those of virgin fibers (VCFs). Similar to this, Yang et al. [18] showed that RCFs may be cleaned with a tiny bit of oxygen in a nitrogen atmosphere without losing their mechanical strength. These results are consistent with more general observations that heat recycling, although efficient, usually degrades the mechanical integrity of recovered fibers, requiring careful processing condition optimization to optimize the possibility for fiber reuse [19].

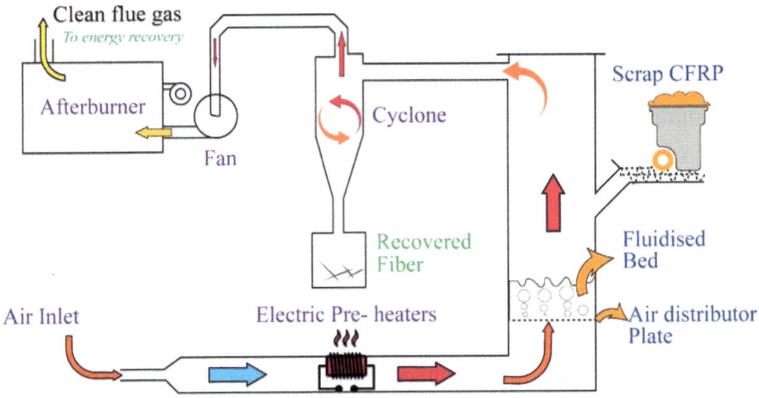

Fig. 8 Thermal recycling through fluidized bed test [6]. Schematic diagram showing the thermal recycling process of materials through a fluidized bed test. The process involves heating materials using heaters in a fluidized bed reactor to break down and recover components of composites for reuse

Fluidized Bed Process: The fluidized bed process is an effective method for recovering high-quality glass fibers (GFs) and carbon fibers (CFs) from end-of-life (EOL) fiber-reinforced polymer composites (FRPCs). This process, which operates at temperatures between 400 and 650 °C, uses a heated silica sand bed fluidized by hot air to decompose the chopped FRPCs as shown in Fig. 8. The thermoset polymer matrix is removed through thermal or oxidative means, resulting in clean fibers. The organic components are vaporized and carried away with the silica particles. Since glass melts at much higher temperatures (1400–1600 °C), the resin provides the necessary heat and energy for the process. The recovered fillers and fibers can then be reused to produce high-performance composites [20].

The recycling of polyester sheet molding compound (SMC) scrap was performed by crushing and sieving it to 25 mm particles, then using a temperature range of 450–600 °C to recover fibers. Recovered glass fibers (RGFs) were washed to remove resin residue and used in new composites with virgin glass fibers (VGFs). The study found that composites with a 50:50 blend of RGFs and VGFs exhibited better flexural and tensile strength, though RGFs retained only 50% of VGFs' strength, suggesting RGFs as a limited alternative. In a study, the parameters for the fluidized bed process were 450 °C and 1.3 m/s velocity to recycle GFs and the recovering fibers were approximately of 5 mm length. Mechanical testing showed a 50% reduction in tensile strength, which dropped by 90% at 600 °C due to fiber deterioration and defects. They recommended lower processing temperatures to preserve strength, finding no major effect on Young's modulus compared to VGFs. The recycled carbon fibers (RCFs) from different types of prepreg were granulated using a step-by-step granulation process with various screen sizes. Smaller screens (6 mm, 8 mm) produced better results. The environmental impact of RCF recycling via a fluidized bed was investigated and found that it required significantly less energy than virgin carbon fibers (VCFs) and reduced life-cycle energy and greenhouse gas emissions by up to 50%, indicating an efficient, sustainable alternative. Table 2 provides a comprehensive overview of various recycling techniques and their outcomes for composite materials, detailing the types of reinforcement fibers, matrix materials, recycling methods, temperatures, energy requirements, and the recovered materials.

4.4 Innovative and Emerging Recycling Techniques

Microwave Assisted Recycling: Compared to conventional thermal processes, microwave-assisted recycling is more selective and energy-efficient since it uses microwave power to heat and decompose thermoset composite. This process targets the polymer matrix while preserving reinforcing fibers as shown in Fig. 9. For epoxy-based carbon fiber reinforced polymers (CFRPs), early successful trials demonstrated effective carbon fiber extraction using microwave irradiation in different atmospheres (Ar, N_2, air). The most effective results were achieved under an argon atmosphere due to spark glow plasma generation. Optimal fiber cleanliness was observed at 500 °C in a nitrogen atmosphere. A recent study introduced a fast, catalyst-free microwave-assisted chemical method using hydrogen peroxide and tartaric acid, achieving significant tensile strength retention in just one minute of microwave exposure. For epoxy-based glass fiber reinforced polymers (GFRPs), a combination of solvolysis and microwave treatment using acetic acid and hydrogen peroxide proved effective. A microwave irradiation of 180 s produced the greatest results, producing recycled glass fibers with mechanical qualities comparable to those of virgin fibers. Glass fibers were also recovered by microwave pyrolysis at 450 °C for an hour, however, some char residue was left behind.

Superheated Steam Recycling: Superheated steam treatment, which involves exposing thermoset composites to high-pressure steam at elevated temperatures, causes the polymer matrix to degrade and allows the fibers to split from the resin. Pure superheated steam is usually used for the first treatment at 400 °C, and then 4% oxygen is added at 500 °C for the second treatment. This technique may be applied as a surface treatment following fiber reclamation and enhances the interfacial adhesion of carbon fibers. Using steam as an oxidant in a pyrolysis furnace at medium temperatures (600–800 °C) is a potential method for recovering carbon fibers from CFRPs. With the exception of chilling, this procedure produces clean fiber surfaces in 140 min. Superheated steam recycling has also been used to successfully recover oligomers from co-polyesters based on terephthalic and ortho-phthalic acid that have been crosslinked with styrene. This procedure has the ability to value co-polyester thermoset composites' reinforcing and matrix components [22].

Electrochemical Recycling: Electrochemical methods apply an electric current to thermoset composites in an electrolyte solution, inducing chemical reactions that break down the polymer chains while preserving reinforcing fibers. This technique shows promise for selectively degrading thermoset materials. For epoxy-based carbon fiber reinforced polymers (CFRPs), electrochemical recycling involves using

Table 2 Recycling techniques and outcomes for composite materials [6]

Reinforcement fibers	Matrix	Technique of recycling	Temperature (°C)	Energy	Recovered material
–	PLA	Mechanical and chemical	–	14 MJ	Matrix
Glass fibers	Elium resin	Mechanical, chemical, and thermal	90, 245, 450	(0.29, 1, 4.20) MJ/Kg	Matrix + fibers
Glass fibers	Polypropylene	Thermal	170–220	–	Fibers
Glass fibers	Unsaturated polyester	Mechanical	400	17.1 to – 89.1 MJ/kg	Fibers + matrix
–	PLA	Mechanical	–	Low energy	Matrix
Carbon fibers	RTM6 epoxy resin	Thermal	320	19 MJ/kg	Fibers + matrix
Carbon fibers	Epoxy Resin	Chemical	160–180	–	Fibers
Basalt fibers	Polypropylene	Thermal	200–600	–	Fibers
Carbon fibers	DER331 epoxy resin	Chemical	190	–	Fibers + matrix
–	Polypropylene	Mechanical + chemical	–	–	Matrix
Glass fibers	Epoxy, polyester, and vinyl ester	Chemical	250–370	–	Fibers + matrix
Glass fibers	PA 66	Mechanical	700	–	Fibers + matrix
Carbon fibers	Epoxy resin	Mechanical	–	2.03 MJ/kg	Fibers + matrix
Carbon fibers	Polar bear epoxy resin	Chemical	180–210	106.6 kW/h	Fibers
Carbon fibers	Bisphenol A-based epoxy resin	Thermal	500	–	Fibers
Glass fibers	Polyester resin	Mechanical and chemical	80	–	Fibers
Hardwood cellulose	High density polyethylene	Mechanical	180–190	–	Cellulose
Carbon fiber	Poly lactic acid	Reverse 3D printing	240	67.7 MJ/kg	Fibers + matrix
Glass fibers	Polypropylene	Mechanical	250	–	Fibers
Glass fibers	Epoxy resin	Thermal	565	–	Fibers
Carbon fibers	Epoxy resin	Chemical	400	Very low energy	Fibers
Glass fibers	Epoxy resin	Thermal	375	12 MJ/g	Fibers + matrix

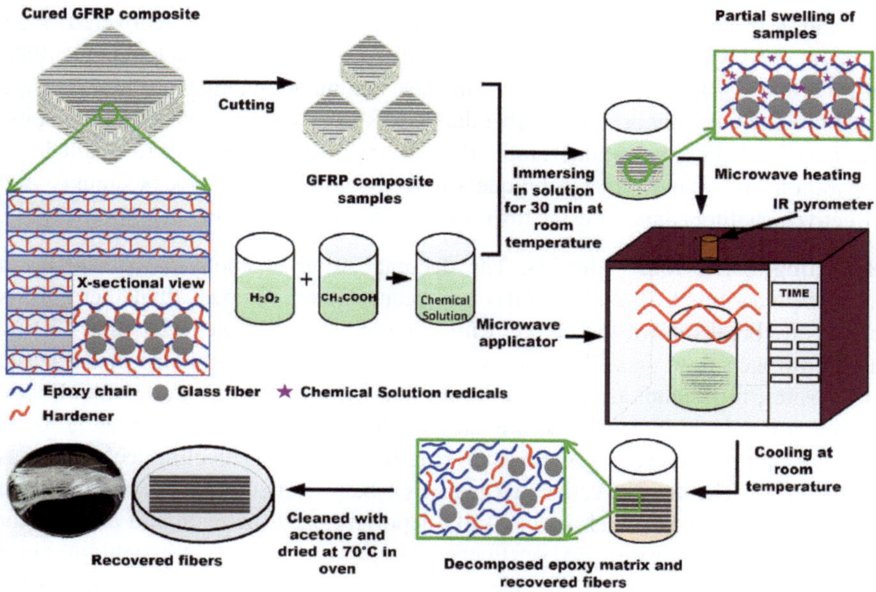

Fig. 9 Microwave-assisted recycling of glass fiber reinforced composite recover glass fibers [21]. The schematic diagram illustrates the microwave-assisted recycling process for glass fiber reinforced composites. The diagram outlines the essential steps, starting with cutting the composite into smaller pieces, which are then immersed in a solvent. Following this, the composite material is heated in a microwave, aiding in the breakdown of the resin matrix. The diagram also depicts the recovery of glass fibers, emphasizing the processes of separation and collection

a direct current power supply with a stainless-steel cathode and the CFRP specimen as the anode in a NaCl solution. Optimal conditions (4 mA current and 3% NaCl) yielded fibers with 80% of their original tensile strength. High-voltage electrical treatments have also been used to remove epoxy resin, with higher electrolyte concentrations improving effectiveness but potentially damaging fibers over long durations. Polyester-based glass fiber reinforced polymers (GFRPs) have been recycled using high-voltage fragmentation, which employs pulse electrical discharges in a dielectric liquid to disintegrate the composite material. This method produces cleaner fibers with longer lengths and lower resin content than mechanical recycling, though it is energy intensive. Continuous carbon fibers from CFRPs can be reclaimed using Joule heating, where the composite is heated to 400 °C to decompose the resin [23].

Ultrasonic Recycling: Through the mechanical disruption of thermoset composites' molecular structure by high-frequency vibrations, ultrasonic recycling breaks down the polymer matrix and makes fiber separation easier. To enhance the recovery of carbon fiber (CF) from CFRP waste, a sonochemical method combining hydrogen peroxide and diluted nitric acid is employed. The two-step process includes pretreatment with an aqueous solution of nitric acid and hydrogen peroxide, followed by sonication in an ultrasonication tank. The tensile strength of the recovered fibers is maintained while resin breakdown is greatly increased using this technique. 98% nitric acid to 2% hydrogen peroxide is the ideal combination ratio since it maximizes resin breakdown without causing fiber damage. Recently, thermoplastic CF composites have been pre-cracked interlaminarly utilizing power ultrasonics to enable layer separation and reuse. This technique initiates-controlled pre-cracks quickly and effectively, without damaging the fibers or compromising mechanical properties [24].

Recycling Using Ionic Liquids: The decomposition of epoxy-based CFRPs in molten potassium hydroxide (KOH) has been studied, showing that thermolysis effectively breaks down the epoxy resin at temperatures between 285 and 330 °C under nitrogen flow. In addition to decomposing different pollutants in the waste composites, this technique preserves more than 95% of the original carbon fibers' tensile strength. Using ionic liquids and alcohols at mild circumstances (150 °C and atmospheric pressure), an environmentally acceptable recycling technique has been created. The most effective mixture was 1-butyl-3-methyl imidazolium acetate with ethylene glycol, which successfully depolymerized the resin and allowed for the recovery and reuse of carbon fibers and epoxy. A recent approach combines solvolysis with ionic liquid degradation, using ethylene glycol and an acid ionic liquid to achieve significant composite disassembly. This method maintains excellent thermal and mechanical properties while achieving up to 95% resin degradation. A new method using sulfuric acid to recycle glass fiber reinforced polymers (GFRPs) was introduced. This process produces polyionic liquids in situ, cleaning the glass fibers and maintaining their properties. As a kind of superplasticizer, the by-product, dissolved epoxy, can be added to cement mixtures to increase compressive strength [25].

Biological Recycling: In order to break down thermoset composites, recent research is looking into biological techniques such microbial treatment and enzymatic breakdown. These methods, though still limited in studies, offer sustainable and environmentally friendly recycling solutions by using enzymes or microorganisms to target specific polymers within the composite matrix. Biotechnology, particularly using oxidative enzymes, shows promise for the biodegradation of complex plastics and composites. For example, a high-throughput screening assay has been developed to identify effective enzymes for breaking down epoxy resins. Applying chemical and mechanical pretreatments to facilitate enzyme access and employing organic solvents to solubilize the materials both help this procedure. Polyurethane matrices, which contain ester and amide bonds that enzymes may break to create useful compounds including alcohols, acids, and aromatic precursors, have been the subject of enzymatic recycling studies. The chemical industry may make use of these recovered resources again. This subject is now being advanced by the EU project Bizente, which involves developing hydrolytic enzymes and oxidoreductases to break down thermoset resins. To recycle thermoset composites and recover resins and other products for future use, this project will develop enzyme-based biodegradation methods [26]. Table 3 outlines the benefits and drawbacks of various composite recycling techniques.

Table 3 Advantages and disadvantages of various composite recycling techniques [19]

Recycling technology	Recovered material/energy	Advantages	Disadvantages
Mechanical	Matrix rich fraction: filler Fiber rich fraction: reinforcement	• Recovery of both fibers and matrix • Absence of hazardous chemical substances • No gas emissions generated • Low energy cost	• Reduction in mechanical properties (including length reduction and degradation) • Inadequate bonding between fiber and matrix • Carbon fiber (CF) causing damage to equipment
Thermal (pyrolysis)	Matrix: energy Fiber: reinforcement	• Fiber length retention • No use of hazardous chemical substances	• Char contamination • Gas emissions • Properties of recovered fibers are significantly influenced by process parameters (with glass fibers particularly affected by high temperatures)
Thermal (fluidized bed)	Matrix: energy Fiber: reinforcement	• High tolerance for contaminated materials (such as paint, foam cores, and metal inserts) • Absence of char contamination	• Reduction in mechanical properties (including length reduction) • No recovery of matrix material • Lower energy recovery compared to pyrolysis • Glass fibers (GF) are adversely affected by high temperatures
Chemical	Matrix turned into oligomers/ monomers Fiber: reinforcement	• Long, clean fibers • Preservation of mechanical properties • Absence of char residue • High recovery rate of the matrix material	• Costly equipment required • Corrosion resistance • Potential use of hazardous solvents • Necessity for post-treatment of solvents • Operation under high temperature and pressure conditions
Microwaved	Matrix: energy Fiber: reinforcement	• Moderate temperature • Reduced damage to glass fibers • No char residue and complete retention of mechanical properties for carbon fibers	• Char residues on glass fibers (GF) • Carbon fibers (CF) develop flaws when exposed to air • Requirement for an argon (Ar) or nitrogen (N_2) atmosphere for carbon fibers (CF)

5 Real-World Examples of RBRFC

5.1 *Automotive Industry*

Recycled composite materials are increasingly found applications in automotive components to minimize the adverse impacts on the environment, increase fuel economy, and decrease weight. Recycled composite components are utilized in a variety of products, including dashboards, door panels, and interior trims as shown in Fig. 10. For example, producers are creating lightweight, outstanding durability carbon fiber prepregs for automobile panels that reduce carbon dioxide emissions and increase fuel effectiveness. Since recycled materials are used for the production so this approach is both scalable and cost-effective. Case studies show how companies have successfully produced exterior body panels with recycled composites to speed up painting processes while lowering production expenses and pollution.

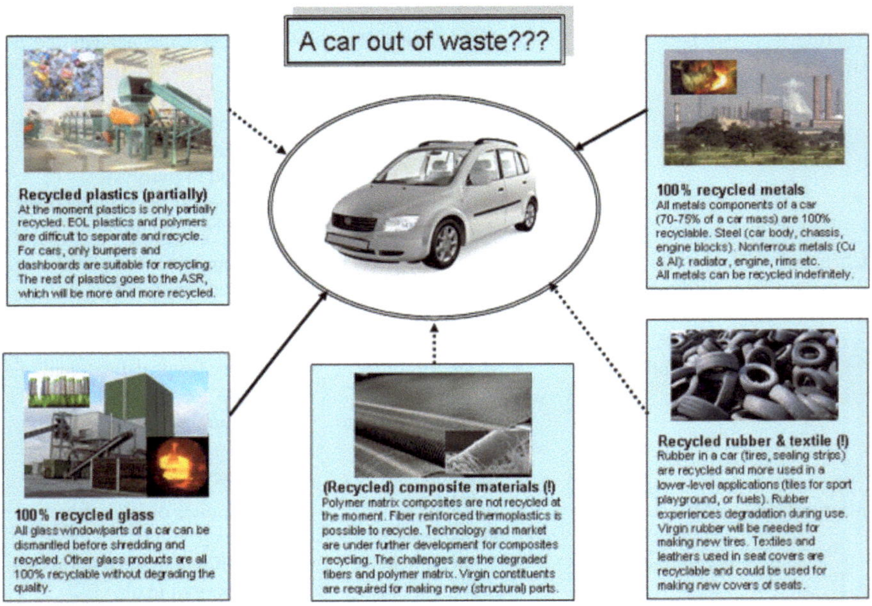

Fig. 10 Possibility of a car made entirely from 100% recycled materials become a reality by 2030, or 2050 [12]. Illustration of a car made entirely from 100% recycled materials, showing the potential reality by 2030 or 2050. Highlights include: 70–75% of car parts are fully recyclable, rubber from tires can be reused in applications like seat covers, and glass components are recyclable. Only the bumpers and dashboards are currently recycled, while other parts are not yet recyclable

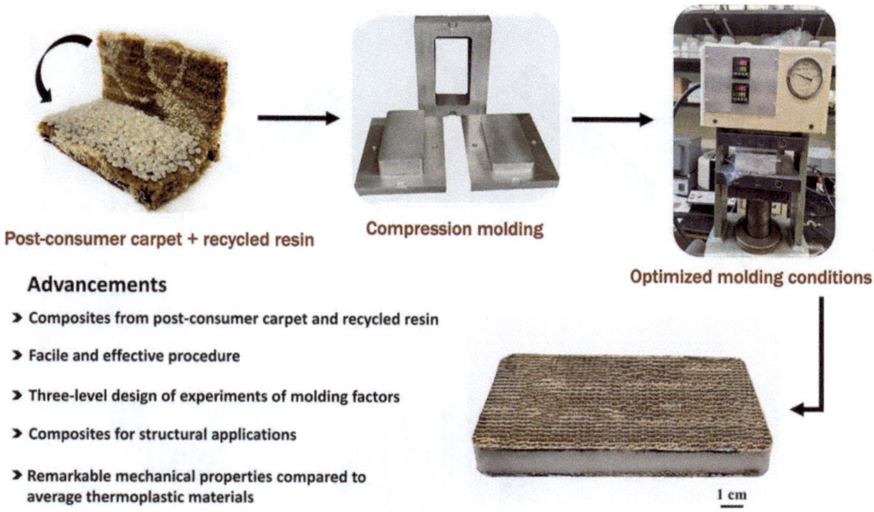

Fig. 11 Schematic diagram illustrating the recycling process of composite materials for structural applications. The process involves feeding post-consumer carpet and recycled resin into a molding machine, followed by optimization to produce a final composite material with enhanced mechanical properties, superior to average thermoplastic materials [27]

5.2 *Construction and Infrastructure*

The construction sector incorporates recycled composites for roofing materials carpeting (Fig. 11), and thermal insulation, among other building components. These materials have several benefits, including increased resilience to moisture, better durability, and enhanced thermal insulation. Sustainable building projects have made use of recycled composites, which are lightweight and robust materials that reduce structural load on structures.

5.3 *Aerospace and Marine Applications*

The marine and aviation industries employ recycled composites because they are robust, lightweight, and corrosion resistant. These materials are helpful for components that reduce repair expenses and improve fuel efficiency, such as aircraft sections, boat hulls, and interior panels for aeroplanes. For instance, recycled carbon fiber composites are utilized in aircraft components to ensure durability and save weight, hence enhancing overall performance as shown in Fig. 12. Marine boats also employ these materials because they have lesser drag, which boosts economy and speed, and they are less likely to corrode in saltwater.

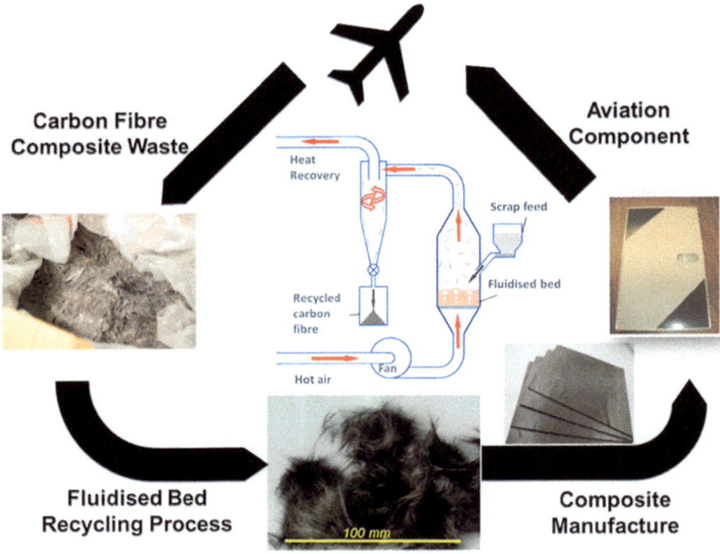

Fig. 12 Schematic diagram showing the process of recycling carbon fiber composite waste from the aviation industry for reuse in new aviation components. The process includes a fluidized bed recycling step, followed by composite manufacturing to produce reusable aviation parts from the recycled material [28]

5.4 Consumer Goods and Electronics

Recycled composites are utilized in consumer goods ranging from electronic casings to furniture. The usage of these materials in products like tables, seats, and shelving units promotes the concepts of the circular economy by lowering waste and resource consumption. Recycled composites are utilized in electronics for cases and parts due to their strength and low weight. Growing consumer acceptance is spurred by increased awareness of environmentally friendly and sustainable products, according to market trends. As more companies look into recycled materials to meet consumer demand for sustainable products, they are starting to appear as a viable alternative across a variety of product categories [12]. Table 4 highlights different methods used for recycling, identifies the companies that use these methods, and provides information about how much recycling capacity these companies have in various countries.

Table 4 A concise overview of various recycling techniques, including the companies involved and their capacities in different countries [6]

Company	Process	Recycling technology	Country	Working capacity
Solvay	Chemical	Prepreg technology	Belgium	–
Hadeg Recycling Ltd	Thermal	Pyrolysis	Germany	–
Wittmann technology GmBH	Mechanical	Granulator	Austria	100–500 ton/year
Alpha Recyclage Composites	Thermal and chemical	Steam thermolysis process	France	300 tons/year
SELFRAG-HVF lab equipment	Chemical	Electrochemical	Switzerland	–
Karborek	Mechanical and thermal	RIFT	Italy	1500 tons/year
CFK Valley Stade Recycling GmbH & Co. KG	Thermal	Pyrolysis	Germany	1000 tons/year
ECRC	Mechanical	Grinding	Denmark, Germany, and France	–
Pilot plant at University of Nottingham	Thermal	Fluidized bed	UK	100 tons/year
MIT-RCF	Thermal and chemical	Pilot-plant pyrolysis	USA	3–5 million lbs/ year
Sigmatex	Mechanical and thermal	Over molding	UK	–
ReFiber ApS	Thermal	Pyrolysis	Denmark	5000 tons/year
JCMA	Mechanical, thermal, and chemical	Crushing, separation, and combustion	Japan	–
Phoenix Fiber glass Inc	Mechanical	Mechanical recycling adopted through chopped and milled fibers	Canada	160 tons/month
ERCOM composite recycling	Mechanical	Crushing and separation is utilized for different particle sizes	Germany	6000 tons/year

6 Regulatory Issues and Standards

6.1 Current Standards for RBRFC

To guarantee performance, quality, and safety, recycled-based reinforced fiber composites (RBRFC) need to meet several requirements. Guidelines on material qualities, testing procedures, and quality assurance for composites are provided by established standards such as ISO and ASTM. To assist producers, convey the recycled content and eco-friendliness of their products, ISO 14021:2016, for example, lays forth rules for environmental labels and declarations. Like this, ASTM standards like ASTM D7209-06 concentrate on assessing the mechanical and physical characteristics of recycled plastic lumber to make sure it satisfies criteria for strength and durability. Whether for the consumer products, construction, or automotive industries, these standards guarantee that RBRFCs fulfil performance requirements unique to their respective industries. Furthermore, safety rules handle certain hazards associated with the usage of composite materials, guaranteeing that recycled elements do not jeopardize the safety of products.

6.2 Environmental Regulations

Policies throughout the world are encouraging more recycling and waste management, which has a direct impact on RBRFC uptake. To promote sustainable material recovery, the European Union's Waste Framework Directive and Circular Economy Action Plan establish high recycling goals and promote the use of recovered materials. RBRFC adoption is strongly impacted by global regulations that encourage greater recycling and waste management. The Circular Economy Action Plan and the Waste Framework Directive of the European Union promote sustainable material recovery by encouraging the use of recycled materials and setting high recycling goals. Strict restrictions on the usage of specific compounds are also imposed by regulatory frameworks in other nations, such as the Registration, Evaluation, Authorization, and Restriction of Chemicals (EU REACH), which further affects the processing and application of recycled materials in the creation of composites.

6.3 Future Trends in Regulatory Landscape

With a stronger focus on sustainability and circular economy principles, the regulatory landscape around RBRFCs is probably going to change. Stricter rules for material traceability are anticipated in future legislation, guaranteeing that recycled parts fulfil stringent quality and safety requirements all the way through the supply chain. Additional incentives, such tax breaks or subsidies for using recycled materials in goods, could also be available to businesses who embrace environmentally friendly methods. Additionally, as the industry develops, new standards that concentrate especially on the special qualities and uses of recycled composites may surface, encouraging their broad use in a variety of industries [29].

7 Strategies for Scaling up Production and Adoption

Strategic actions in a number of sectors are needed to increase the production and integration of recycled composites. The development of sophisticated recycling methods that effectively treat durable composite materials is a crucial aspect of technological innovation. For instance, high-quality recycled feedstock may be produced using specialized shredders and granulators. Recycled composites are also being included using large-scale additive manufacturing methods, which can improve overall material performance and assist control thermal characteristics. Utilizing bio-based materials that enhance sustainability and mechanical qualities, such cellulose, flax, and hemp, is the main goal of material development. These natural fibers are frequently recyclable and improve composites' life cycle evaluation. Additionally, hybrid composites that include virgin and recycled fibers preserve performance standards while lessening their negative effects on the environment. Cost optimization techniques, such as adding sensors to recycling machinery to track and boost operational effectiveness, which lowers maintenance and downtime, increase economic feasibility. Companies may be persuaded to use recycled composites in their manufacturing processes by financial incentives like as tax rebates and subsidies. With regulations encouraging the use of recycled materials in sectors like manufacturing and building, regulatory support is important. By ensuring that recycled composites fulfil quality and safety standards, standards and certification programs boost industry trust and promote wider use. Growth requires industry collaboration since industry consortia assist share resources, expertise, and best practices, reducing the risks involved with adopting new materials, and relationships with research institutions may propel developments in sustainable manufacturing technology. The usage of recycled composites can be de-risked and creative solutions can result from collaborations with institutions such as Oak Ridge National Laboratory [30].

References

1. Pickering SJ (2006) Recycling technologies for thermoset composite materials—current status. Compos Part A Appl Sci Manuf 37:1206–1215. https://doi.org/10.1016/j.compositesa.2005.05.030
2. Khurshid MF, Hengstermann M, Hasan MMB, Abdkader A, Cherif C (2020) Recent developments in the processing of waste carbon fibre for thermoplastic composites—a review. J Compos Mater 54:1925–1944. https://doi.org/10.1177/0021998319886043

3. Kim K-W, Kim D-K, Han W, Kim B-J (2022) Comparison of the characteristics of recycled carbon fibers/polymer composites by different recycling techniques. Molecules 27:5663. https://doi.org/10.3390/molecules27175663
4. Shehab E, Meiirbekov A, Amantayeva A, Tokbolat S (2022) Cost modelling for recycling fiber-reinforced composites: state-of-the-art and future research. Polymers (Basel). 15:150. https://doi.org/10.3390/polym15010150
5. Asmatulu E, Twomey J, Overcash M (2014) Recycling of fiber-reinforced composites and direct structural composite recycling concept. J Compos Mater 48:593–608. https://doi.org/10.1177/0021998313476325
6. Khalid MY, Arif ZU, Ahmed W, Arshad H (2022) Recent trends in recycling and reusing techniques of different plastic polymers and their composite materials. Sustain Mater Technol 31:e00382. https://doi.org/10.1016/j.susmat.2021.e00382
7. Schinner G, Brandt J, Richter H (1996) Recycling carbon-fiber-reinforced thermoplastic composites. J Thermoplast Compos Mater 9:239–245. https://doi.org/10.1177/089270579600900302
8. Palmer J, Savage L, Ghita OR, Evans KE (2010) Sheet moulding compound (SMC) from carbon fibre recyclate. Compos Part A Appl Sci Manuf 41:1232–1237. https://doi.org/10.1016/j.compositesa.2010.05.005
9. Bledzki A, Spaude R, Ehrenstein GW (1985) Corrosion phenomena in glass fibers and glass fiber reinforced thermosetting resins. Compos Sci Technol 23:263–285. https://doi.org/10.1016/0266-3538(85)90040-5
10. Torres A (2000) Recycling by pyrolysis of thermoset composites: characteristics of the liquid and gaseous fuels obtained. Fuel 79:897–902. https://doi.org/10.1016/S0016-2361(99)00220-3
11. Piñero-Hernanz R, García-Serna J, Dodds C, Hyde J, Poliakoff M, Cocero MJ, Kingman S, Pickering S, Lester E (2008) Chemical recycling of carbon fibre composites using alcohols under subcritical and supercritical conditions. J Supercrit Fluids 46:83–92. https://doi.org/10.1016/j.supflu.2008.02.008
12. Yang Y, Boom R, Irion B, van Heerden D-J, Kuiper P, de Wit H (2012) Recycling of composite materials. Chem Eng Process Process Intensif 51:53–68. https://doi.org/10.1016/j.cep.2011.09.007
13. Bai Y, Wang Z, Feng L (2010) Chemical recycling of carbon fibers reinforced epoxy resin composites in oxygen in supercritical water. Mater Des 31:999–1002. https://doi.org/10.1016/j.matdes.2009.07.057
14. Okajima I, Hiramatsu M, Sako T (2011) Recycling of carbon fiber reinforced plastics using subcritical water. Adv Mater Res 222:243–246. https://doi.org/10.4028/www.scientific.net/AMR.222.243
15. Jiang J, Deng G, Chen X, Gao X, Guo Q, Xu C, Zhou L (2017) On the successful chemical recycling of carbon fiber/epoxy resin composites under the mild condition. Compos Sci Technol 151:243–251. https://doi.org/10.1016/j.compscitech.2017.08.007
16. Zhu J-H, Chen P, Su M, Pei C, Xing F (2019) Recycling of carbon fibre reinforced plastics by electrically driven heterogeneous catalytic degradation of epoxy resin. Green Chem 21:1635–1647. https://doi.org/10.1039/C8GC03672A
17. Meyer LO, Schulte K, Grove-Nielsen E (2009) CFRP-recycling following a pyrolysis route: process optimization and potentials. J Compos Mater 43:1121–1132. https://doi.org/10.1177/0021998308097737
18. Yang J, Liu J, Liu W, Wang J, Tang T (2015) Recycling of carbon fibre reinforced epoxy resin composites under various oxygen concentrations in nitrogen–oxygen atmosphere. J Anal Appl Pyrolysis 112:253–261. https://doi.org/10.1016/j.jaap.2015.01.017
19. Bernatas R, Dagreou S, Despax-Ferreres A, Barasinski A (2021) Recycling of fiber reinforced composites with a focus on thermoplastic composites. Clean Eng Technol 5:100272. https://doi.org/10.1016/j.clet.2021.100272
20. Mastellone ML, Perugini F, Ponte M, Arena U (2002) Fluidized bed pyrolysis of a recycled polyethylene. Polym Degrad Stab 76:479–487. https://doi.org/10.1016/S0141-3910(02)00052-6

21. Rani M, Choudhary P, Krishnan V, Zafar S (2022) Development of sustainable microwave-based approach to recover glass fibers for wind turbine blades composite waste. Resour Conserv Recycl 179:106107. https://doi.org/10.1016/j.resconrec.2021.106107
22. Jiang L, Ulven CA, Gutschmidt D, Anderson M, Balo S, Lee M, Vigness J (2015) Recycling carbon fiber composites using microwave irradiation: reinforcement study of the recycled fiber in new composites. J Appl Polym Sci 132. https://doi.org/10.1002/app.42658
23. Sun H, Guo G, Memon SA, Xu W, Zhang Q, Zhu J-H, Xing F (2015) Recycling of carbon fibers from carbon fiber reinforced polymer using electrochemical method. Compos Part A Appl Sci Manuf 78:10–17. https://doi.org/10.1016/j.compositesa.2015.07.015
24. Ragupathi B, Bacher MF, Balle F (2023) First efforts on recovery of thermoplastic composites at low temperatures by power ultrasonics. Clean Mater 8:100186. https://doi.org/10.1016/j.clema.2023.100186
25. Pérez RL, Ayala CE, Opiri MM, Ezzir A, Li G, Warner IM (2021) Recycling thermoset epoxy resin using alkyl-methyl-imidazolium ionic liquids as green solvents. ACS Appl Polym Mater 3:5588–5595. https://doi.org/10.1021/acsapm.1c00896
26. Podara C, Termine S, Modestou M, Semitekolos D, Tsirogiannis C, Karamitrou M, Trompeta A-F, Milickovic TK, Charitidis C (2024) Recent trends of recycling and upcycling of polymers and composites: a comprehensive review. Recycling 9:37. https://doi.org/10.3390/recycling9030037
27. Azarfam MY, Maheshwari A, Blum FD, Chaudhari S, Switzer C, Vaidyanathan R, Hanan JC, Bandla S (2024) Recycled carpet-reinforced composites from post-consumer polypropylene carpet and recycled HDPE resin. Resour Conserv Recycl 200:107298. https://doi.org/10.1016/j.resconrec.2023.107298
28. Meng F, Cui Y, Pickering S, McKechnie J (2020) From aviation to aviation: Environmental and financial viability of closed-loop recycling of carbon fibre composite. Compos Part B Eng 200:108362. https://doi.org/10.1016/j.compositesb.2020.108362
29. Minkov N, Lehmann A, Winter L, Finkbeiner M (2020) Characterization of environmental labels beyond the criteria of ISO 14020 series. Int J Life Cycle Assess 25:840–855. https://doi.org/10.1007/s11367-019-01596-9
30. Composite world (2024) Scaling up sustainable solutions for fiber-reinforced composites. https://www.compositesworld.com/articles/scaling-up-sustainable-solutions-for-materials-of-the-future-2

Recycled Regenerated Fiber-Reinforced Composites (RFRC) and Their Applications

Sana Ullah

Abstract In the realm of composite manufacturing, natural fibers have gained popularity due to their sustainability advantages. Plant-based biofibers are increasingly replacing synthetic fibers in composites, leading to the development of various biocomposites. Each type of natural fiber has unique chemical compositions and properties, requiring detailed analysis for comparison. Extensive research has highlighted the reinforcement potential of natural fibers, resulting in the production of high-performance biocomposites after years of study. Specifically, plant fibers like bast and leaf fibers are extensively utilized in the automotive sector. However, many other fibers, often examined only in laboratory settings, have yet to achieve large-scale commercial success. Additionally, it's important to consider fibers derived from seeds (such as coir) and animal sources (like chicken feathers), as these are secondary products or waste materials. While bast fibers are frequently reviewed, other plants and animal fibers receive less attention. This review explores six types of plant fibers—bast, leaf, seed, straw, grass, and wood alongside animal fibers and regenerated cellulose fibers. It also delves into recent advancements in natural fibers and their composites, discussing fiber sources, extraction methods, availability, composition, and mechanical properties. The review outlines the benefits and drawbacks of each biofiber and briefly addresses three fabric architectures: nonwoven, woven, and knitted. Lastly, this chapter provides an overview of composite results from such fibers, with references for further detailed studies.

1 Introduction

The global fiber production in 2023 rose to 124 million tons which was just 38 million tons in 2009, that significant increase in just 15 years also raised concerns of recycling and reusing the huge amount of textile waste. Synthetic fibers have the highest share in the produced textiles, i.e., more than 70% of the total fibers are

S. Ullah (✉)
Institute for Fiber Engineering and Science (IFES), Interdisciplinary Cluster for Cutting Edge Research (IFES), Shinshu University, Ueda, Japan
e-mail: sana_ullah@shinshu-u.ac.jp

H. Jamshaid and U. Zubair (eds.), *Sustainable Recycling of Fiber Reinforced Polymer Composites*, SDGs and Textiles, https://doi.org/10.1007/978-981-96-9536-2_8

based on synthetic materials (*Source* The Global Fiber Market, Materials Market report 2024). The increasing production of synthetic fibers is also impacting the increase in textile waste which ultimately affects the climate and economy at large. Although the production of natural fibers is less than that of synthetic ones, there is still no significant post-consumer treatment of the textile waste from either source. The escalating demand for sustainable materials has led to significant advancements in the field of composite materials, particularly in the development of Recycled Regenerated Fiber-Reinforced Composites (RFRC). RFRCs, which incorporate fibers reclaimed from post-consumer or post-industrial waste, offer a promising solution to both environmental concerns and the need for high-performance materials in various industries.

This chapter delves into the current state of RFRC technology, focusing on the types of regenerated fibers used, the manufacturing processes, and the applications across different industries. The discussion also includes challenges in the commercialization of RFRCs and the future directions of this technology.

Regenerated fibers can be obtained from various sources; regenerated protein is one of the sustainable sources for the conversion of biomass to usable products [1]. Poole et al. explained very briefly the circle of eco-friendly or sustainable fibers. Figure 1 represents the generally accepted concept of eco-friendly fiber production. The term eco-friendly is widely used for recycling and reusing the green/sustainable materials. However, the process of recycling may involve some steps which are not supported by the claim of eco-friendliness. Only such processes which include renewable and biodegradable materials combined with green processing should result the product which are fully sustainable. Regenerated fiber-reinforced composites (RFRC) are also one of the sustainable materials if the process from collection of biomass waste to regeneration of fibers reinforced composites undergoes the same eco-friendly methodology.

Reusing plants or biomass-derived organic fibers promotes responsible waste management and a circular economy. Although raw material is still bio waste here, it is processed to obtain fibers with consistent properties such as viscose rayon, and Tencel.

This chapter underlines the possible regenerated fiber-reinforced composites (RFRC) and their utilization in various industries. This chapter further points out how RFRC is already contributing toward sustainability. The possibilities of recycling and utilization of the post-consumer waste coming from RFRC would be addressed. The technologies, techniques, sustainability implications, challenges, and opportunities in the recycling of RFRC are being addressed in this chapter. Finally, the future perspective of this category of composites will be highlighted.

Recycling of Regenerated Wool Fibers

Wool keratin is one of the most sustainable sources of fibers, however, recycling wool has been a huge challenge for researchers [2]. Li and Wang suggested an eco-friendly way of recycling the wool keratin and regeneration of films based on recycled wool keratin. They proposed the use of ionic liquid solutions for the preparation of regenerated wool films, a significant addition to their claim of eco-friendly recycling

Fig. 1 Concept of being eco-friendly in fiber industry

of the textile biomass [3]. Similarly, Poole et al. also suggested the environmentally sustainable textiles from the recycled proteins (regenerated keratin). Figure 2 represents a schematic illustration of production of regenerated keratin from various biomass sources, and then converting it to the fibers by textile processing techniques [1].

Recycled Carbon Fibers

Carbon fiber having exceptional strength to weight ratio, is one of the most demanding materials for aerospace, automotive, and renewable energy. In the context of global economy and textile market, carbon fiber reinforced composites possess huge contributions. Currently the carbon fiber composite market is more than 20 billion USD (in 2023, Precedence Research), that shows the interest of fiber industry to produce carbon fiber to fulfill the high demand. Recycling carbon fibers is a crucial step toward sustainable material management, especially given the increasing use of carbon fiber-reinforced polymers (CFRP) in industries like aerospace, automotive,

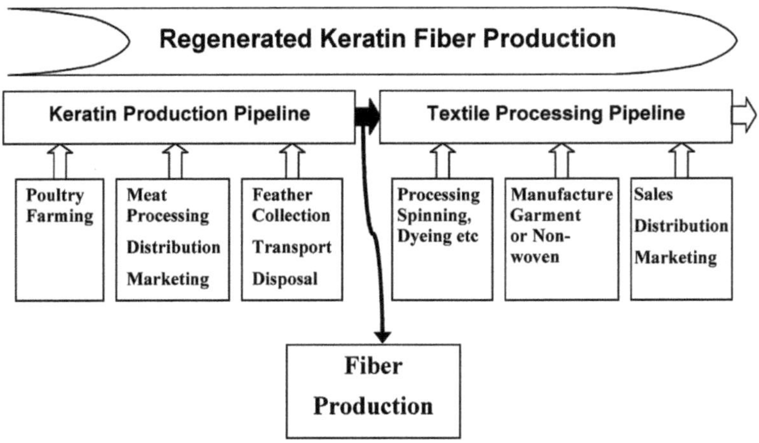

Fig. 2 Production of regenerated keratin fiber "Reprinted (adapted) with permission from [1]. Copyright 2009, American Chemical Society"

and sports equipment [4]. The recycling process typically involves breaking down the composite material to recover the carbon fibers, which can then be reused in new products. One common method is pyrolysis [5], where the composite is heated in the absence of oxygen to decompose the resin matrix, leaving behind the carbon fibers. Another method is solvolysis, which uses solvents to dissolve the resin, preserving the integrity of the carbon fibers. Recent advancements include microwave thermolysis, which is faster and more energy-efficient than traditional methods. The recovered carbon fibers can be used to create new composite materials, reducing waste and the demand for virgin carbon fibers. This not only helps in managing waste but also contributes to the circular economy by extending the lifecycle of carbon fibers [4].

Regenerated Cellulose Fibers

Cellulose is the most abundant organic polymer on Earth, derived from sources like wood pulp, bamboo, and cotton linters. This makes it a sustainable and readily available raw material. Regenerated cellulose fibers can be processed into various forms, such as viscose, lyocell, and modal, each with unique properties suitable for different applications. The manufacturing process allows for the adjustment of fiber properties, such as length, diameter, and strength, to meet specific requirements. Regenerated cellulose fibers are biocompatible, making them suitable for medical applications like wound dressings and surgical materials [6]. Regenerated cellulose fibers are biodegradable, reducing environmental impact and promoting sustainability. They have excellent moisture-absorbing capabilities, providing comfort in textiles and hygiene products. These fibers exhibit high tensile strength and modulus (as shown in Fig. 3) [7], enhancing durability and resistance to wear and tear. They offer superior breathability, making them ideal for clothing and other textile applications. Using regenerated cellulose fibers can reduce reliance on virgin cotton and decrease textile waste, contributing to environmental conservation for future apparel. Regenerated

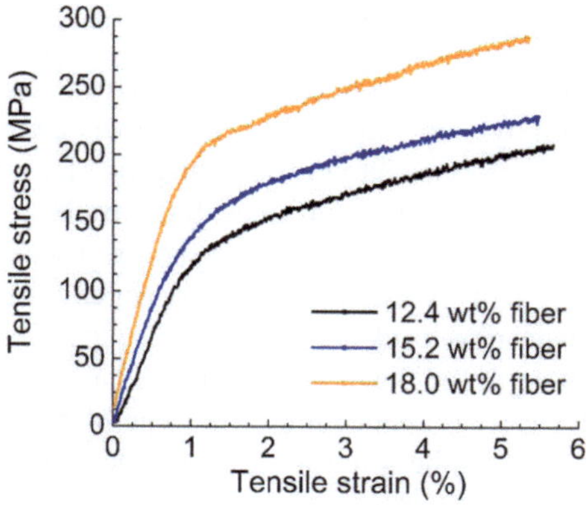

Fig. 3 Typical trend of Young's modulus of regenerated cellulose fibers "Reprinted (adapted) with permission from [7] Copyright 2016, American Chemical Society"

cellulose fibers combine the benefits of natural and synthetic materials, making them indispensable in various industries, from fashion to healthcare [6]. A recent investigation (2024) by Zeng et al. also found that regenerated cellulose fibers (RCF) have high mechanical properties and they also proposed the use of regenerated cellulose fibers in temperature adaptive thermal management [8].

Recycled regenerated cellulose fibers exhibit quite good strength comparable to commercially available viscose. Even secondary recycled cellulose fibers have considerably fair strength, a recent study shows that secondary RCFs exhibit only 14.6% lesser strength as compared to primary CRCF. Recycling of colored regenerated cellulose from cotton waste contributes hugely toward the reduction of textile waste and plays a vital role in the sustainability [9]. Recycling cotton waste requires special treatment as recycling may affect color retention, strength, elongation at break, and other factors. Therefor pretreatments are done before the recycling of cellulose [10]

Ever increasing demand of energy poses pressure on the energy generation and storage sectors. RCF can be one of the useful and sustainable sources for energy storage, especially for the application of wearable electronics. Cellulose fibers lack conductivity, which can be imparted through blending carbon fibers in regenerated cellulose hydrogels, forming a composite material which possess the required feature [11]. Cellulose fibers are not only in high demand in technical fields, but the fashion industry is also majorly dependent on the regenerated cellulose fibers. A survey was conducted on the sustainability of future apparel in the U.S, where 65% of the total participants (544) expressed a willingness to purchase regenerated cellulose fiber made clothing considering the environmental implications and climate change [12].

Despite having excellent functional properties, the processing of RCF has been a great challenge. Dissolution and spinning of RCF is a complex process; however, there have been recent studies presenting the ideas to improve the spinnability of the regenerated cellulose. Tu et al. proposed to use alkali/urea system to produce the spun fibers from regenerated cellulose [13].

Regenerated Silk Fibers

Regenerated silk fibers are derived from natural silk, typically from the cocoons of the Bombyx mori silkworm. These fibers are processed to extract silk fibroin, which is then reformed into various materials. Here are some key points about regenerated silk fibers and their composites: Silk fibroin is highly biocompatible, making it suitable for biomedical applications such as tissue engineering and drug delivery systems. These fibers are biodegradable, which is beneficial for environmental sustainability. Regenerated silk fibers exhibit excellent mechanical properties, including high tensile strength and flexibility. Silk fibers have good thermal stability, making them suitable for various applications. They can be processed into different forms, such as films, gels, and fibers, allowing for a wide range of applications. Silk composites are used in tissue engineering, wound dressings, and surgical sutures due to their biocompatibility and biodegradability. Regenerated silk fibers are used in the textile industry for making high-quality fabrics with desirable tactile properties [14]. Electronics: Silk-based composites are explored for use in flexible electronics and sensors due to their mechanical strength and electrical properties. Environmental Applications: These composites are used in filtration systems and environmental sensors due to their biodegradability and functional properties. Recent advancements have led to the development of silk bio-composites that combine silk fibers with other natural or synthetic materials. These composites exhibit improved mechanical characteristics, biocompatibility, and sustainability. For example, magnetic artificial silk fibers (Fig. 4) have been developed for use in soft robotics and biomedical devices [15].

Regenerated Fiber-Reinforced Composites (RFRCs)

The term "composite" refers to any material that combines two or more different substances to create something new. Think of it like a sandwich—each layer (or ingredient) brings its own unique flavor and texture to the final product. In the world of materials, these layers are held together by adhesives, binders, or chemical bonds.

Composite materials are born from blending two or more substances with distinct physical or chemical properties. When these materials come together, the resulting composite has its own unique set of characteristics, different from those of its individual components. Figure 5 represents a schematic view of a composite structure.

Here are the key players in composite material:

Matrix: This is the continuous phase that surrounds and binds the reinforcement material, distributing stress and protecting it. Imagine it as the bread holding everything together in our sandwich analogy. Matrices can be made of polymers, metals, ceramics, or even carbon.

Reinforcement: This is the stronger, stiffer part that provides most of the composite's mechanical properties. It's like the hearty fillings in our sandwich. Common reinforcement materials include fibers (like carbon or glass), particles, or flakes. The magic happens when these components work together, resulting in composites with amazing properties, such as high strength-to-weight ratio: Composites can be much

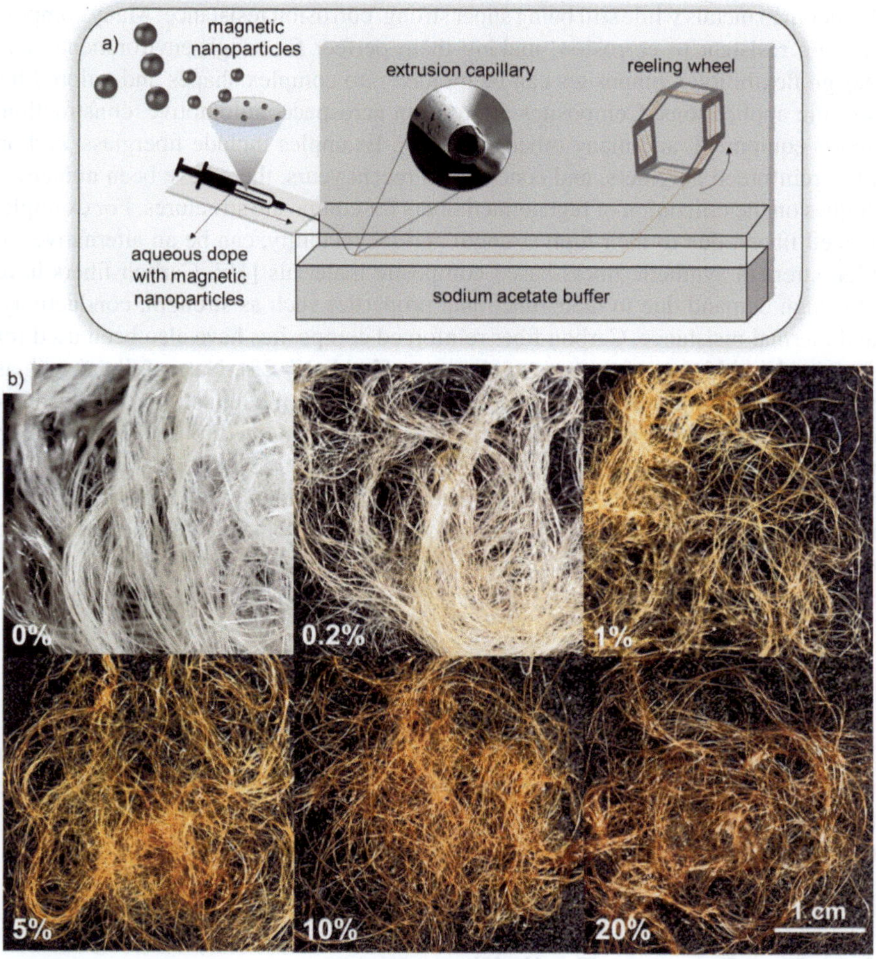

Fig. 4 Spinning of the magnetic artificial silk fibers. **a** Schematic of the spinning setup. The nanoparticles were mixed with the recombinant spider silk protein solution (300 mg/mL in Tris–HCl pH 8) to generate spinning dopes containing 0.2–20% w/w magnetite, **b** magnetic silk with 0–20% w/w of magnetic nanoparticles [15]

Fig. 5 General schematic of a composite structure

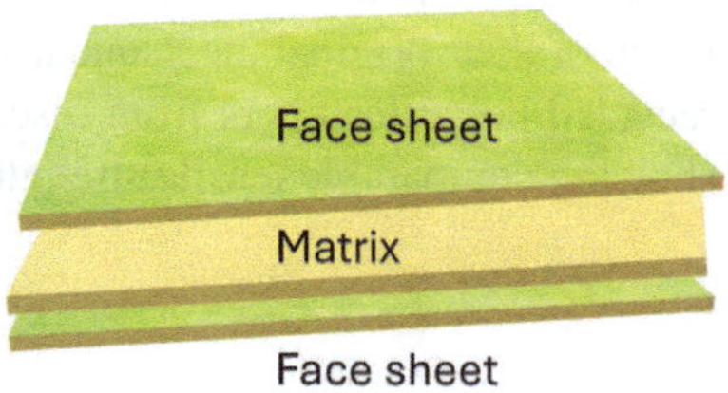

lighter than metals while still being super strong, corrosion resistance: Many composites are resistant to corrosion, making them perfect for tough environments, and design flexibility: Composites can be molded into complex shapes and tailored for specific applications. Composites are used in aerospace, automotive, construction, sports equipment, and many other industries. Examples include fiberglass, carbon fiber-reinforced polymers, and concrete. In recent years, there have been numerous studies on the utilization of regenerated fibers for composite structures. For example, ioncell fibers, due to their high strength and recyclability, can be an alternative for high strength synthetic fibers-based composite materials [16]. Carbon fibers have also high demand due to their functional properties such as strength, conductivity, and thermal resistance. Carbon fiber reinforced composites have also been used for structural and high-performance applications [5, 11, 17–22]. Natural fibers such as wool, cotton, and jute are sustainable sources for the textile industry; however, recycling such natural fibers is quite challenging. There is a need to pay consideration to the recycling of natural fibers and blending it with recycled polymer resins to form natural fibers-based composite materials, which not only fulfill the demand and required parameters for desired applications but also contribute to the sustainable development of the industry [2, 6, 23–31].

2 Types of Recycled Regenerated Fibers

Recycled regenerated fibers are primarily derived from three sources: post-consumer textile waste, industrial waste, and natural fibers. Each source contributes diverse types of fibers, which can be used independently or blended to create RFRCs with tailored properties.

2.1 Post-Consumer Textile Waste

Post-consumer textile waste, such as discarded garments and household textiles, is a rich source of fibers like cotton, polyester, and nylon. These fibers can be reclaimed through mechanical or chemical processes, resulting in regenerated fibers that retain much of their original mechanical strength. A recent study demonstrated the use of recycled polyester fibers from discarded PET bottles in the production of RFRCs, showing comparable tensile strength to virgin fiber composites [32, 33].

2.2 Industrial Waste

Fibers recovered from industrial processes, such as carbon fiber trimmings from aerospace manufacturing, are another valuable source of reinforcement for composites. These fibers are often shorter and may have surface defects, but they can be treated to enhance their bonding with the matrix material. The incorporation of recycled carbon fibers from aerospace waste into polypropylene matrices has been shown to improve the composite's mechanical properties, particularly its stiffness and thermal resistance [34, 35].

2.3 Natural Fibers

Natural fibers such as jute, hemp, and flax are increasingly being recycled and regenerated into composite materials. These fibers are biodegradable, renewable, and can be sourced from agricultural waste, making them highly sustainable. Researchers have explored the use of recycled jute fibers in polymer composites, finding that they offer good mechanical properties while reducing the overall environmental impact of the material [36, 37].

3 Manufacturing Processes for RFRCs

The manufacturing of RFRCs involves a combination of traditional composite fabrication techniques and novel methods tailored to the challenges posed by recycled fibers.

3.1 Mechanical Recycling and Fiber Reclamation

The mechanical recycling process involves shredding or grinding the waste material to reclaim fibers. This process is often used for thermoplastic composites, where the matrix can be melted and reformed around the recycled fibers. A study highlighted the effectiveness of mechanical recycling in producing RFRCs from carbon fiber waste, with the resultant material showing a 15% increase in impact resistance compared to virgin composites [38, 39].

3.2 Chemical Recycling and Regeneration

Chemical recycling processes involve breaking down the polymer matrix to release the embedded fibers. These fibers can then be reprocessed into new composites. This method is particularly effective for thermosetting composites, where the matrix cannot be easily remelted. The chemical recycling of glass fiber composites from wind turbine blades has been successfully demonstrated, with the regenerated fibers being reused in new composite materials for automotive applications [20, 40].

3.3 Hybrid Techniques

Hybrid techniques combine mechanical and chemical recycling processes to maximize fiber recovery and enhance the quality of the regenerated fibers. These methods are especially useful for dealing with mixed-material waste streams. A recent innovation in hybrid recycling integrates pyrolysis with mechanical fiber extraction, resulting in high-purity carbon fibers from mixed waste.

4 Applications of RFRCs

The versatility and sustainability of RFRCs have led to their adoption across various industries, including automotive, construction, aerospace, and consumer goods.

4.1 Automotive Industry

The automotive industry is a significant consumer of RFRCs, using them for interior panels, bumper reinforcements, and under-the-hood components. The use of RFRCs not only reduces the weight of vehicles, leading to better fuel efficiency, but also contributes to the circular economy by reusing waste materials. The integration of RFRCs in car door panels, made from recycled carbon fibers and polypropylene, has been shown to reduce vehicle weight by up to 20% without compromising safety.

4.2 Construction Industry

In construction, RFRCs are used for load-bearing structures, insulation panels, and even in concrete reinforcement. Their durability and resistance to environmental degradation make them ideal for use in harsh conditions. Recycled glass fiber composites have been employed in the reinforcement of bridge decks, providing a lightweight and corrosion-resistant alternative to traditional steel reinforcements.

4.3 Aerospace Industry

While the aerospace industry primarily uses high-performance virgin composites, there is a growing interest in RFRCs for secondary structures, such as non-critical panels and interior components. The use of RFRCs in these applications reduces material costs and supports sustainability goals. Boeing has begun exploring the use of RFRCs for aircraft interior panels, with preliminary tests indicating that these materials meet the necessary flame retardancy and mechanical strength requirements.

4.4 Consumer Goods

The consumer goods sector has seen a surge in the use of RFRCs, particularly in the production of durable goods such as furniture, sporting equipment, and electronic casings. The aesthetic flexibility of RFRCs allows for the creation of products that are both functional and visually appealing. A startup has developed a line of eco-friendly furniture using RFRCs made from recycled denim fibers and biopolymer matrices, offering a sustainable alternative to traditional wood and plastic furniture.

5 Challenges in RFRC Development

Despite their advantages, RFRCs face several challenges that need to be addressed for their wider adoption.

5.1 Variability in Fiber Quality

The quality of recycled fiber can vary significantly depending on their source and the recycling process used. This variability can lead to inconsistencies in the mechanical properties of RFRCs, making it difficult to guarantee performance. Studies have shown that the mechanical properties of RFRCs made from recycled textile fibers can vary by as much as 30%, depending on the degree of fiber degradation during recycling. Treatment conditions also impact the physical properties of final fibers, such as heat treatment may cause loss in mechanical strength as shown in Fig. 6. It was observed that during the heat treatment of recycled glass fiber, the strength decreased with the increase of treatment temperature. The decline in mechanical strength was

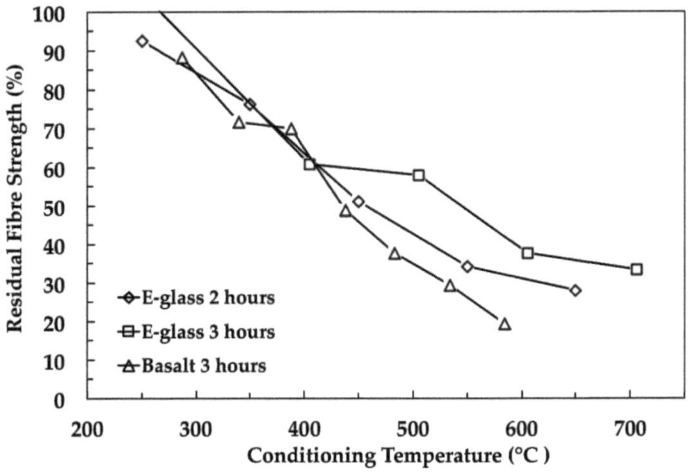

Fig. 6 Typical results from the literature on residual fiber strength after heat conditioning [41]

irrespective of the type of glass [41]. In a result of these types of treatment, the recycled fibers will also influence the final strength of composite materials. However, there are also some techniques by which the strength of recycled fiber can be regenerated. A detailed study of regeneration of recycled glass fibers have been carried out by Thomason et al., where they proposed the use of hot sodium hydroxide to regain the lost strength of glass fiber. Ku et al. carried out an investigation of carbon fiber reinforced composite and were able to get recycled composite with almost 100% recyclability [42]. Depending on the treatment method, recycling technique, treatment conditions, and types of fibers/fabric, there may be variation in the resultant recycled regenerated fibers.

5.2 Compatibility with Matrices

Recycled fibers often have surface contaminants or degraded properties that affect their bonding with the polymer matrix. This issue can be mitigated through surface treatments or the use of coupling agents, but these add complexity and cost to the manufacturing process. The use of silane coupling agents has been shown to improve the interfacial bonding between recycled glass fibers and epoxy matrices, resulting in composites with enhanced tensile strength.

5.3 *Economic Viability*

The economic viability of RFRCs is still a concern, particularly in comparison with virgin composites. While the cost of recycled fibers is lower, the additional processing required to ensure consistent quality can offset these savings. A cost analysis of RFRCs made from recycled carbon fibers found that, while raw material costs were 20% lower, the overall cost savings were only 5% after accounting for processing and quality control.

6 Future Directions and Opportunities

The future of RFRCs lies in overcoming the current challenges and expanding their applications. Research and development efforts are focusing on improving the consistency and quality of recycled fibers, as well as exploring new recycling technologies that can handle a broader range of waste materials. Innovations in recycling technologies, such as enzymatic recycling and supercritical fluid processing, have the potential to improve the quality of regenerated fibers and expand the types of materials that can be recycled. Enzymatic recycling of polyester textiles has been shown to produce high-purity fibers that can be directly reused in the production of RFRCs, offering a more sustainable alternative to traditional chemical recycling methods. As RFRC technology matures, its applications are expected to expand into new areas, such as medical devices, renewable energy, and high-performance sports equipment. These sectors require materials that are not only strong and durable but also lightweight and sustainable. Researchers are exploring the use of RFRCs in wind turbine blades, where the reduced weight and increased sustainability of the material could lead to significant improvements in energy efficiency and lifecycle impact. In recent decades, technology has exponential growth, combining modern technologies with sustainable materials would lead to advancements in the RFRC. Nanotechnology plays a key role in providing high surface area and lightweight materials, covering a considerable part of the textile science and technology. Composite fibers can be produced not only by conventional spinning techniques but also by electrospinning and similar techniques to produce fibers having diameters in nanoscale range [43, 44]. Electrospun composite nanofibers containing regenerated polymers are also becoming popular amid escalation in climate change [45–47]. Electrospinning of biomaterials such as deoxyribonucleic acid (DNA) [48], zein [49], gelatin [50], hemp [51], lignin [52], and similar polymers is challenging but there have been several studies expressing the potential of regenerated and sustainable fibers using biomaterials as a core or support matrices.

7 Conclusion

Recycled Regenerated Fiber-Reinforced Composites represent a significant step forward in the pursuit of sustainable, high-performance materials. While there are challenges to be addressed, the potential benefits of RFRCs in terms of environmental impact, material performance, and economic viability make them a promising area of research and development. As industries continue to prioritize sustainability, the demand for RFRCs is likely to grow, driving further innovation and adoption of these materials across a wide range of applications. RFRC can also be trialed for nanoscale fabrication and advanced technical applications such as regenerated fiber based electrospun nanofibrous composite materials for biomedical, environmental, energy conversion and storage, food preservation, and many other as scope of RFRCs is expanding.

Author Contributions As a sole author, all contributions are attributed to the same.

Data Availability There was no data used in this research. Figures/tables derived from the other sources have been acknowledged and permissions were taken prior to use. Microsoft embedded copilot was used for reefing the English language of the manuscript.

Conflict of Interest The author declares no conflict of interest.

References

1. Poole AJ, Church JS, Huson MG (2009) Environmentally sustainable fibers from regenerated protein. Biomacromol 10:1–8
2. Mu B, Yu X, Yang Y (2023) Sustainable and green process for recycling waste wool textiles into high-quality protein fibers on a pilot scale. Resour Conserv Recycl. https://doi.org/10.1016/j.resconrec.2023.107190
3. Li R, Wang D (2013) Preparation of regenerated wool keratin films from wool keratin-ionic liquid solutions. J Appl Polym Sci 127:2648–2653. https://doi.org/10.1002/app.37527
4. Chen PY, Feng R, Xu Y, Zhu JH (2023) Recycling and reutilization of waste carbon fiber reinforced plastics: current status and prospects. Polymers (Basel) 15
5. Deng J, Xu L, Zhang L et al (2019) Recycling of carbon fibers from CFRP waste by microwave thermolysis. Processes. https://doi.org/10.3390/pr7040207
6. Varnaitė-Žuravliova S, Baltušnikaitė-Guzaitienė J (2024) Properties, production, and recycling of regenerated cellulose fibers: special medical applications. J Funct Biomater 15:348. https://doi.org/10.3390/jfb15110348
7. Zhu C, Richardson RM, Potter KD et al (2016) High modulus regenerated cellulose fibers spun from a low molecular weight microcrystalline cellulose solution. ACS Sustain Chem Eng 4:4545–4553. https://doi.org/10.1021/acssuschemeng.6b00555
8. Zeng C, Wang H, Bu F et al (2024) A regenerated cellulose fiber with high mechanical properties for temperature-adaptive thermal management. Int J Biol Macromol. https://doi.org/10.1016/j.ijbiomac.2024.133550
9. Liu H, Fan W, Miao Y et al (2023) Closed-loop recycling of colored regenerated cellulose fibers from the dyed cotton textile waste. Cellulose 30:2597–2610. https://doi.org/10.1007/s10570-022-05002-w

10. Ma Y, Rosson L, Wang X, Byrne N (2020) Upcycling of waste textiles into regenerated cellulose fibres: impact of pretreatments. J Text Inst 111:630–638. https://doi.org/10.1080/00405000.2019.1656355
11. Carvalho JT, Cunha I, Coelho J et al (2022) Carbon-yarn-based supercapacitors with in situ regenerated cellulose hydrogel for sustainable wearable electronics. ACS Appl Energy Mater 5:11987–11996. https://doi.org/10.1021/acsaem.2c01222
12. Chi T, Frattali A, Liu H, Chen Y (2023) Regenerated cellulose fibers (RCFs) for future apparel sustainability: insights from the U.S. consumers. Sustainability (Switzerland) 15. https://doi.org/10.3390/su15065404
13. Tu H, Li X, Liu Y et al (2022) Recent progress in regenerated cellulose-based fibers from alkali/urea system via spinning process. Carbohydr Polym 296
14. Anulaya SV, Kandasubramanian B (2025) Advancements in silk bio-composites for multi-faceted applications. In: Kandasubramanian B, Jaya Prakash N (eds) Engineering natural silk: applications and future directions. Springer Nature Singapore, Singapore, pp 53–70
15. Greco G, Schmuck B, Del Bianco L et al (2024) High-performance magnetic artificial silk fibers produced by a scalable and eco-friendly production method. Adv Compos Hybrid Mater. https://doi.org/10.1007/s42114-024-00962-y
16. Fazeli M, Islam S, Baniasadi H et al (2024) Exploring the potential of regenerated Ioncell fiber composites: a sustainable alternative for high-strength applications. Green Chem 26:6822–6835. https://doi.org/10.1039/d3gc03637e
17. Ateeq M (2023) A state of art review on recycling and remanufacturing of the carbon fiber from carbon fiber polymer composite. Compos Part C: Open Access. https://doi.org/10.1016/j.jcomc.2023.100412
18. Feng H, Xu X, Wang B et al (2023) Facile preparation, closed-loop recycling of multifunctional carbon fiber reinforced polymer composites. Compos B Eng. https://doi.org/10.1016/j.compositesb.2023.110677
19. Nistratov AV, Klimenko NN, Pustynnikov IV, Vu LK (2022) Thermal regeneration and reuse of carbon and glass fibers from waste composites. Emerg Sci J 6:967–984. https://doi.org/10.28991/ESJ-2022-06-05-04
20. Karuppannan Gopalraj S, Kärki T (2020) A review on the recycling of waste carbon fibre/glass fibre-reinforced composites: fibre recovery, properties and life-cycle analysis. SN Appl Sci 2
21. Shen MY, Guo ZH, Feng WT (2023) A study on the characteristics and thermal properties of modified regenerated carbon fiber reinforced thermoplastic composite recycled from waste wind turbine blade spar. Compos B Eng. https://doi.org/10.1016/j.compositesb.2023.110878
22. Chen CH, Chiang CL, Wang JX, Shen MY (2022) A circular economy study on the characterization and thermal properties of thermoplastic composite created using regenerated carbon fiber recycled from waste thermoset CFRP bicycle part as reinforcement. Compos Sci Technol. https://doi.org/10.1016/j.compscitech.2022.109761
23. Karmakar GP (2022) Regeneration and recovery of plastics. In: Encyclopedia of materials: plastics and polymers. Elsevier, pp 634–651
24. Zhao X, Copenhaver K, Wang L et al (2022) Recycling of natural fiber composites: challenges and opportunities. Resour Conserv Recycl 177
25. Xu X, Tian F, Li X (2020) Rubber-filled wood fiber compos
26. Kim T, Kim D, Park Y (2022) Recent progress in regenerated fibers for "green" textile products. J Clean Prod 376
27. Cao GS, Zhang ZP, Rong MZ, Zhang MQ (2022) A novel strategy for producing high-performance continuous regenerated fibers with wool-like structure. SusMat 2:90–103. https://doi.org/10.1002/sus2.46
28. Elsayed S, Hellsten S, Guizani C et al (2020) Recycling of superbase-based ionic liquid solvents for the production of textile-grade regenerated cellulose fibers in the lyocell process. ACS Sustain Chem Eng 8:14217–14227. https://doi.org/10.1021/acssuschemeng.0c05330
29. Biswas MC, Dwyer R, Jimenez J et al (2021) Strengthening regenerated cellulose fibers sourced from recycled cotton T-shirt using glucaric acid for antiplasticization. Polysaccharides 2:138–153. https://doi.org/10.3390/polysaccharides2010010

30. Wang L, Huang S, Wang Y (2022) Recycling of waste cotton textile containing elastane fibers through dissolution and regeneration. Membranes (Basel). https://doi.org/10.3390/membranes12040355
31. Qiao T, Yang C, Zhao L et al (2024) High tensile regenerated cellulose fibers via cyclic freeze-thawing enabled dissolution in phosphoric acid for textile-to-textile recycling of waste cotton fabrics. Int J Biol Macromol. https://doi.org/10.1016/j.ijbiomac.2024.133911
32. Signorini C, Volpini V (2021) Mechanical performance of fiber reinforced cement composites including fully-recycled plastic fibers. Fibers 9:1–15. https://doi.org/10.3390/fib9030016
33. Bin Bakri MK, Rahman MR, Yurkin Y et al (2022) Recycled plastic and textile waste biocomposites. In: Rahman MR, Bin Bakri MK (eds) Recycled plastic biocomposites. Woodhead Publishing, pp 97–118
34. Heidarian P, Mokhtari F, Naebe M et al (2024) Reclamation and reformatting of waste carbon fibers: a paradigm shift towards sustainable waste management. Resour Conserv Recycl 203
35. Fernández A, Santangelo-Muro M, Fernández-Blázquez JP et al (2021) Processing and properties of long recycled-carbon-fibre reinforced polypropylene. Compos B Eng. https://doi.org/10.1016/j.compositesb.2021.108653
36. Sanvezzo PB, Branciforti MC (2021) Recycling of industrial waste based on jute fiber-polypropylene: manufacture of sustainable fiber-reinforced polymer composites and their characterization before and after accelerated aging. Ind Crops Prod. https://doi.org/10.1016/j.indcrop.2021.113568
37. Benhamadouche L, Rokbi M, Osmani H et al (2021) Characterization of physical and mechanical properties of recycled jute fabric reinforced polypropylene composites. Polym Compos 42:5435–5444. https://doi.org/10.1002/pc.26235
38. Kumar A, Dixit S, Singh S et al (2024) Recent developments in the mechanical properties and recycling of fiber-reinforced polymer composites. Polym Compos
39. Gopalraj SK, Kärki T (2020) A study to investigate the mechanical properties of recycled carbon fibre/glass fibre-reinforced epoxy composites using a novel thermal recycling process. Processes. https://doi.org/10.3390/PR8080954
40. Thomason JL, Nagel U, Yang L, Sáez E (2016) Regenerating the strength of thermally recycled glass fibres using hot sodium hydroxide. Compos Part A Appl Sci Manuf 87:220–227. https://doi.org/10.1016/j.compositesa.2016.05.003
41. Thomason J, Jenkins P, Yang L (2016) Glass fibre strength-a review with relation to composite recycling. Fibers 4
42. Yu K, Shi Q, Dunn ML et al (2016) Carbon fiber reinforced thermoset composite with near 100% recyclability. Adv Funct Mater 26:6098–6106. https://doi.org/10.1002/adfm.201602056
43. Ullah S, Hashmi M, Lee J et al (2022) Recent advances in pre-harvest, post-harvest, intelligent, smart, active, and multifunctional food packaging. Fibers Polym 23:2063–2074. https://doi.org/10.1007/s12221-022-4192-2
44. Parın FN, Ullah S, Yıldırım K et al (2021) Fabrication and characterization of electrospun folic acid/hybrid fibers: in vitro controlled release study and cytocompatibility assays. Polymers (Basel). https://doi.org/10.3390/polym13203594
45. Hashmi M, Ullah S, Ullah A et al (2021) Carboxymethyl cellulose (CMC) based electrospun composite nanofiber mats for food packaging. Polymers 13:302. https://doi.org/10.3390/POLYM13020302
46. Hashmi M, Ullah S, Ullah A et al (2020) Optimized loading of carboxymethyl cellulose (CMC) in tri-component electrospun nanofibers having uniform morphology. Polymers 12:2524. https://doi.org/10.3390/POLYM12112524
47. Khan MQ, Kharaghani D, Sanaullah et al (2019) Fabrication of antibacterial electrospun cellulose acetate/silver-sulfadiazine nanofibers composites for wound dressings applications. Polym Test 74:39–44. https://doi.org/10.1016/j.polymertesting.2018.12.015
48. Ullah S, Ali HG, Hashmi M et al (2023) Electrospun composite nanofibers of deoxyribonucleic acid and polylactic acid for skincare applications. J Biomed Mater Res A 111:1798–1807. https://doi.org/10.1002/jbm.a.37592

49. Ullah S, Hashmi M, Khan MQ et al (2019) Silver sulfadiazine loaded zein nanofiber mats as a novel wound dressing. RSC Adv 9:268–277. https://doi.org/10.1039/C8RA09082C
50. Ullah S, Hashmi M, Shi J, Kim IS (2023) Fabrication of electrospun PVA/Zein/gelatin based active packaging for quality maintenance of different food items. Polymers (Basel). https://doi.org/10.3390/polym15112538
51. Ullah S, Ohsawa O, Ishaq T et al (2023) Fabrication of novel hemp charcoal nanofiber membrane for effectual adsorption of heavy metal ions from wastewater. Sustainability 15:9365. https://doi.org/10.3390/su15129365
52. Haider MK, Kharaghani D, Sun L et al (2023) Synthesized bioactive lignin nanoparticles/polycaprolactone nanofibers: a novel nanobiocomposite for bone tissue engineering. Biomaterials Advances 144:213203. https://doi.org/10.1016/j.bioadv.2022.213203

Life Cycle Assessment of Fiber-Reinforced Polymer Composites for Evaluating Sustainability in Recycling

Usman Zubair

Abstract Fiber-reinforced polymer (FRP) composites are widely employed in industries such as automotive, aerospace, wind energy, and construction owing to their superior strength-to-weight ratio and robustness. However, their ecological footprint, driven by energy-intensive manufacturing and end-of-life disposal challenges, demands sustainable recycling strategies. This chapter explores the life cycle assessment (LCA) of FRP composites, focusing on the environmental impact of raw material extraction, manufacturing, in-use performance, and end-of-life scenarios. It examines mechanical, thermal, and chemical recycling methods, comparing their environmental trade-offs through LCA studies. Key challenges included here are data gaps, complex multi-material compositions, durability trade-offs, and lack of standardized methodologies. Case studies from diverse industries such as automotive, wind energy, and construction sectors highlight real-world recycling applications and circular economy approaches. The chapter also considers emerging policies, industry best practices, and innovations in recyclable resins and fiber recovery to enhance sustainability. By evaluating different recycling pathways, this chapter presents insights into reducing waste, cutting carbon emissions, and fostering a circular economy for FRP composites. The findings here aim to guide researchers, industry stakeholders, and policymakers in selecting efficient and sustainable FRP recycling solutions.

1 Introduction

Fibers-reinforced polymer composites (FRP) have appeared as a promising material for engineering structural applications owing to their high strength-to-weight ratio, durability, and versatility. However, these composite structures' sustainability and end-of-life (EoL) utilization have raised numerous concerns. The recycling of FRP is quite challenging due to the intermixing of diverse materials that occur in

U. Zubair (✉)
Department of Materials, School of Engineering and Technology, National Textile University, Faisalabad, Pakistan
e-mail: usman.zubair@ntu.edu.pk

H. Jamshaid and U. Zubair (eds.), *Sustainable Recycling of Fiber Reinforced Polymer Composites*, SDGs and Textiles, https://doi.org/10.1007/978-981-96-9536-2_9

different forms within the structure. FRP composites are replacing metal-based structures in the construction, aerospace, automotive, renewable energy, and engineering industries at one end. Composites perform better in their service life and require less maintenance credit goes to better resistance to environmental weathering and no rusting that are frequently attributed to metallic structures. On the other end, composites are swapping the wood materials in sports, leisure, furniture, and construction because of their better strength, dexterity, and performance. The matrices used in the manufacturing of FRP composites are either thermoset or thermoplastic (Fig. 1). Although thermoset resins have been used commonly, now thermoplastic matrices are preferred as they offer a better opportunity for recycling the composites. Similarly, bio-based resins or vitrimers (resins based on dynamic covalent bonds [1]) are making their way into composites. The sustainability in FRP is also being addressed by exploring the natural fibers coming from agricultural wastes instead of synthetic fibers. Synthetic fibers offer better properties in terms of mechanical performance, but all the applications do not require composites with greater strength. Figure 2 presents all possible fibers used as reinforcement in the manufacturing of FRP. Advanced fiber manufacturing techniques such as automated fiber placement and additive manufacturing allow the realization of FRP with minimal waste and precise product design for complex structures. The longevity and reliability of composites have now been met through performance optimization tools such as structural health monitoring and damage predictive models. All these advancements in composite manufacturing have allowed us to achieve more sustainable products. As composite materials are making their way in every field of life, their sustainability standing is still being questioned due to poor end-of-life management.

There is a dire need to underline the environmental impact assessment of composite materials through life cycle assessment (LCA). It is also important to determine the recent trends in composite recycling and end-of-life (EoL) reutilization strategies that will extend the life cycle, lower the carbon footprint, and lessen the environmental impact of the composite materials. Owing to the use of diverse materials from various resources, LCA of FRP composites is a challenging activity. Recent

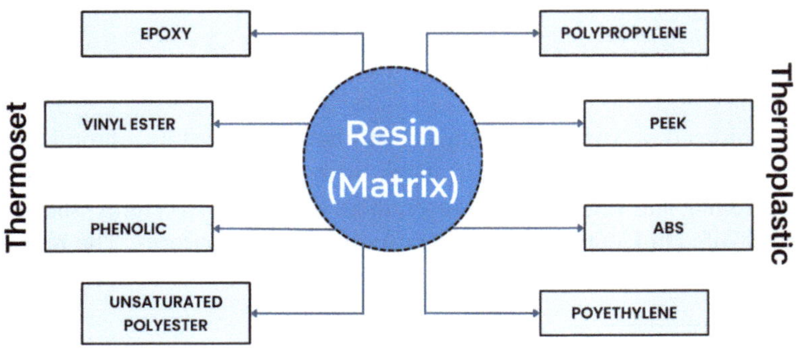

Fig. 1 Resin matrices used for the construction of FRP composites

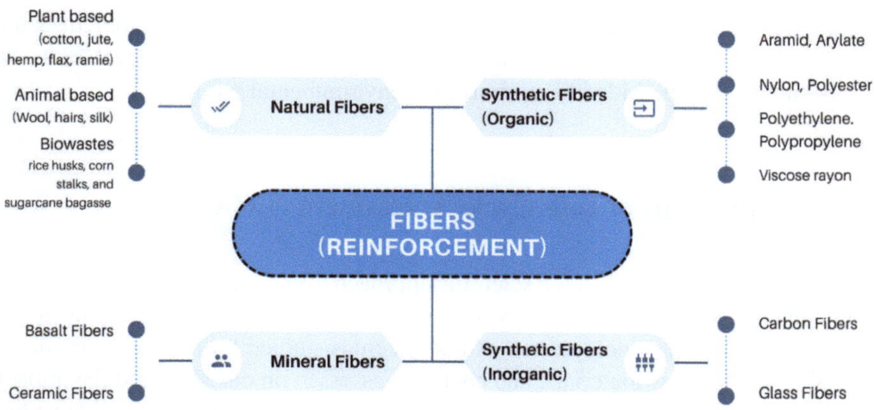

Fig. 2 Fibers as reinforcements used for the construction of FRP composites

advances such as switching from non-recyclable thermosets to recyclable and renewable matrices are also needed to undergo LCA to gauge compliance with circular economic principles. The environmental impact of composites can be minimized by opting for the right recycling technologies despite landfilling or incineration. LCA is an optimal approach to gauge environmental impacts and explore eco-friendly practices both in materials and manufacturing [2].

This chapter will report the strategies to extend the life cycle of FRP composites with a special focus on EoL management and recycling solutions. This document will cover topics on including environmental concerns regarding raw materials extraction, production, and disposal, EoL challenges associated with conventional landfilling and incineration, recycling technologies and their comparative LCA, industrial case studies from the automotive and renewable energy sectors, and future direction and policy recommendations. In this chapter, the comprehensive evaluation of the challenges and opportunities linked to the sustainability and recycling of FRP composites will be addressed. LCA allows the assessment of the environmental impacts of different recycling approaches such as thermal, mechanical, and chemical against conventional landfill and incineration approaches. It allows to specify the carbon footprints, energy consumption, and resource depletion associated with composite manufacturing and EoL management. This chapter will further explore the technical feasibility, economic viability, and environmental trade-offs and underline the barriers and challenges such as high processing costs, materials degradation, and lack of large-scale recycling infrastructure linked to different recycling technologies. The sustainability and circularity among the composite materials can also be realized by extending the lifespan through reuse, repurposing, refurbishment, and downcycling. Circular economic practices encompass both open-loop and closed-loop recycling models. Various case studies will be discussed in this context. Comparative or scenario-based LCA approaches allow us to compare different recycling methods for real-time assessment of energy efficiency, and environment friendliness. This chapter will also underscore the role of governmental policies and regulation, extended producer responsibility (EPR) framework, and best industrial practices for recycling FRP composites. The recommendations for the design of recyclability of composites will be provided to minimize the environmental burden.

2 Fundamentals of Life Cycle Assessment (LCA)

Life cycle assessment (LCA) is a scientific approach to estimating the potential environmental impact and hazard of a product or a service over its complete life cycle. In this methodology, each unit process of the system is studied for the consumption of natural resources, raw materials, and energy is assessed on one side, and the impact of waste streams in the form of gaseous emissions, liquid effluents, and solid residue on the other side [3]. The first step of LCA is the determination of inventories of all the inputs and outputs of each phase of the life cycle in the form of materials and energy that are either flowing into the system or released to the environment. The collected data is then used to assess the environmental impact of the system. The ecological impact is indicated through several impact categories depending on the selected method. The LCA permits comparing different materials and identifying the pollution transference from one impact category to another or one life cycle stage to another (also known as burden shifting). Similarly, this comparison and identification can be carried out between the two different states of the same system, or between two different systems. There are different approaches to performing LCA of a system including but not limited to cradle-to-grave vs. cradle-to-gate, attributional vs. consequential LCA, comparative LCA, Scenario analysis, and circular economy approach (Fig. 3). LCA is used in the framework of "design for the environment" or as support for the decision-making regarding the suitability of the product. Whatever the approach is, LCA needs to be conducted according to the requisites of International Standards. International Standard Organization (ISO) established two frameworks including ISO 14040 (Environmental Management—Life cycle assessment—Principles and Framework) and ISO 14044 (Environmental Management—Life cycle assessment—Requirements and guidelines). A complete LCA, consistent with ISO standards, is comprised of four interrelated phases (Fig. 4).

2.1 Goal and Scope Definition

The first stage of the LCA framework is the goal and scope definition to designate and describe the product or process under investigation. The other activities included in this stage are the establishment of the context in which the assessment is to be conducted and the identification of the system boundaries and ecological impact indicators to be reviewed for the assessment. An important aspect of this phase of LCA is the designation of a functional unit that is the measure of the performance

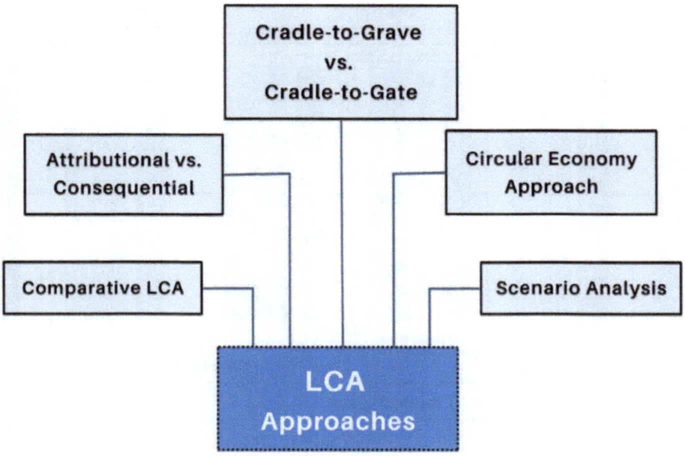

Fig. 3 Different approaches to performing LCA

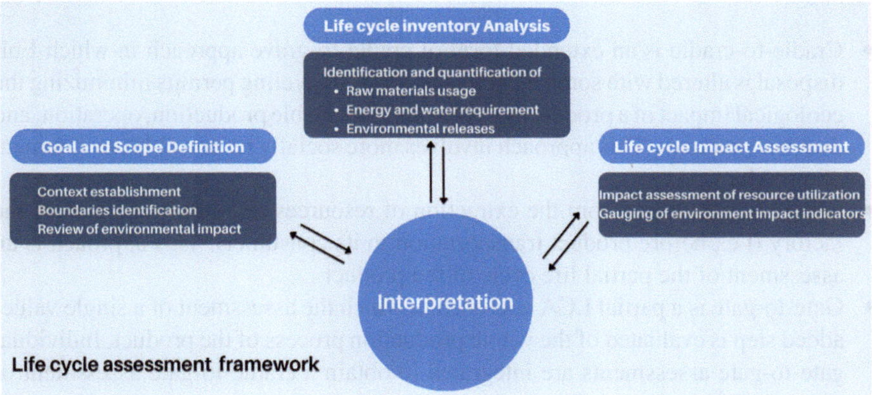

Fig. 4 Stages of life cycle assessment (LCA) framework as per ISO 14044

of the studied system, product, or process. The objective of the functional unit is to establish a reference to which all inputs and outputs are related.

Another context within the goal and scope definition phase is to determine the system boundaries. System boundary outlines the criteria that regulate which unit processes, inputs, outputs, and impacts to be considered in an LCA activity. Four distinct system boundaries can be differentiated as presented in Fig. 5.

- Cradle-to-grave is the full form of LCA starting from the mining, harvesting, and extraction of raw materials (cradle) to the disposal of composites (grave) upon the completion of service life. The two most common disposal approaches are landfilling and incineration.

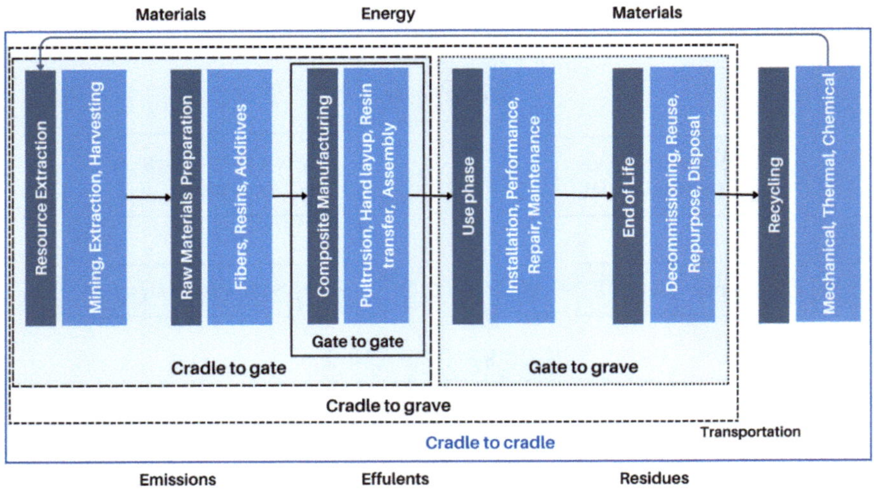

Fig. 5 Different system boundaries for performing LCA

- Cradle-to-cradle is an extended form of cradle-to-grave approach in which EoL disposal is altered with some recycling process. Recycling permits minimizing the ecological impact of a product by deploying sustainable production, operation, and disposal practices. This approach involves more socially responsible development of a product.
- Cradle-to-gate starts from the extraction of resources (cradle) to the gate of the factory (i.e., before product transportation to its consumer). This approach is an assessment of the partial life cycle of the product.
- Gate-to-gate is a partial LCA approach in which the assessment of a single value-added step is evaluated of the whole production process of the product. Individual gate-to-gate assessments are integrated to obtain a cradle-to-gate assessment of the entire production chain.

2.2 *Life Cycle Inventory Analysis (LCIA)*

This step helps identify and quantify energy, water, and raw material usage at one end and environmental releases (e.g., gaseous emissions, solid waste disposal, wastewater effluents) on the other end. It is a technical process of collecting data to quantify the system's inputs and outputs. There are several challenges in performing the inventory analysis of composites that are discussed in subsequent sections.

2.3 Life Cycle Impact Assessment

In this stage, the potential effects of energy, water, and raw material usage have been assessed. Additionally, the ecological impacts of environmental releases that have been gauged in the inventory analysis stage are determined. Results are described through several environmental impact indicators, such as human toxicity, climate change, fossil fuel consumption, ozone depletion, cumulative energy demand, and eutrophication. The carbon footprint of the process is also presented through carbon-related indicators including carbon dioxide equivalent (CO_2-eq.). Impact analysis is also equally challenging for composites manufacturing processes.

2.4 Interpretation

Lastly, the results are reported for interpretation and critically reviewed, however, this stage is interlinked with all three previous stages to comprehend the uncertainties and assumptions made to accomplish reliable assessment. In this step, the results of life cycle inventory analysis and impact assessment are evaluated to arrive at the sustainably favored product, process, or service.

2.5 Challenges in Employing LCA to Composites

Employment of LCA to FRP composites especially involving the recycling stage is quite challenging due to limited availability and poor quality of the data, complex manufacturing technologies, long and uncertain service life, diverse recycling pathways, challenges in environmental impact allocation, and lack of standardized methodologies. Here, all those challenges will be addressed in the LCA framework of the FRP composites. One of the major challenges in the deployment of LCA to the composite is the nonexistence of standardized databases leads to an estimation of unreliable environmental impact. Due to the great variability in materials (fiber type, resin chemistry, and nature of additives) used to construct FRP composites, it becomes challenging to standardize LCA datasets. Composite manufacturing is a complex process and is highly energy-intensive process owing to high energy demands in the stages of fibers manufacturing, composite fabrication, and resin curing. FRP composites can be manufactured by diverse manufacturing techniques such as resin transfer molding, filament winding, spray lay-up, pultrusion, and autoclave curing, which complicates the assessment and induces emissions and waste generation variations.

FRP composites have long service lives, and their robustness complicates the assessment of under-use degradation, maintenance, and replacement cycles. These uncertainties in the life cycle and the EoL scenarios (such as landfilling, incineration, mechanical, thermal, and chemical recycling) make it difficult to develop long-term LCA models. Emerging recycling approaches may lead to closed-loop recycling, but with existing technologies recycled fibers lose their strength and make them only suitable for low-end applications. To gauge the performance of the secondary material from composites' recycling against virgin materials makes LCA studies difficult in the prospect of circularity. The deployment of diverse materials such as metals, coatings, and adhesives along with FRP composites complicates the environmental impact assessment. Issues of impact allocation also surfaced in the absence of EoL scenario as energy recovery bears different environmental credit to recycled fibers recovery. The lack of standardized LCA methodologies for composites also confounds the process. The practiced LCA standards do not fully address the composite-specific challenges. Owing to non-standard practice in composites, the difference in the system boundaries, functional units, and impact categories lead to inconsistent outcomes across the studies. In short, employment of LCA is a challenging task for the impact assessment in the FRP composite industry.

3 Environmental Impact of FRP Composites Across Their Life Cycle

The extraction and processing of raw materials for FRP composites including fibers, resins, and additives, substantially contribute to carbon emissions, pollution, energy consumption, and resource depletion. Similarly, the manufacturing and processing of FRP composites offer critical environmental challenges owing to high energy consumption, emissions, and waste generation. Carbon fiber production remains the most energy-intensive stage, while resin processing contributes to VOC emissions and hazardous waste. By adopting renewable energy sources, emission control technologies, and recycling initiatives, industries can reduce the environmental footprint of FRP manufacturing. While glass and carbon fibers remain dominant, natural fibers offer a more sustainable and favorable alternative. Similarly, bio-based resins bestow an opportunity to lessen fossil fuel dependence and toxicity. Thermoplastic FRP composites also offer the needed advantages of being reused and recycled easily. A combination of recycling, renewable and sustainable materials, cleaner production techniques, and process optimization is essential to mitigating the environmental impact of FRP composite manufacturing [4]. Subsequent sections outline the environmental impacts of fibers, resins, and additives.

The in-use phase of FRP composites offers both environmental benefits and sustainability challenges. While their high durability, corrosion resistance, and lightweight properties contribute to energy efficiency and long service life, their limited recyclability and high embodied carbon profess long-term concerns. A systemic approach is required to reach a balance between durability and sustainability, involving a combination of improved material circularity, advanced recycling technologies, and LCA-driven design strategies. The EoL management of FRP composites also poses significant environmental challenges, mainly due to landfills, incineration emissions, and the loss of valuable materials. Current dumping techniques undermine sustainability efforts by generating GHG emissions, pollution, and resource inefficiencies. These challenges can be addressed through a multi-pronged approach, including advancing recycling technologies, developing bio-based and thermoplastic alternatives [5], enforcing stricter regulations, and adopting circular economy principles. By opting for sustainable disposal and recycling strategies, industries can alleviate the environmental burden of FRP waste, and participate in a more resource-efficient future.

3.1 Raw Material Extraction and Processing (Fibers, Resins, Additives).

Fiber-reinforced polymer (FRP) composites are widely used in aerospace, automotive, construction, wind energy, and marine industries due to their lightweight properties, high strength, corrosion resistance, and durability. However, the sustainability of FRP materials is increasingly scrutinized due to the significant environmental impact associated with raw material extraction and processing. The production of FRP involves the use of fibers (such as glass and carbon), polymer resins (such as epoxy and polyester), and various additives. Each of these components contributes to resource depletion, energy consumption, greenhouse gas (GHG) emissions, and pollution.

A Life Cycle Assessment (LCA) of FRP raw materials provides insights into their environmental footprint, helping industries make informed decisions regarding sustainable sourcing, material efficiency, and circular economy approaches. This chapter explores the environmental impact of extracting and processing fibers, resins, and additives, highlighting key challenges and potential mitigation strategies.

3.1.1 Environmental Impact of Fiber Production

Fibers, as reinforcement, play a critical role in defining the mechanical properties, performance, and durability of FRP composites. The two most extensively used fibers, glass fibers, and carbon fibers, have distinct production processes, each with unique environmental challenges. Glass fibers (GF) are produced from silica (SiO_2) as a base raw material coming from sand that is mixed with network modifiers such as limestone, dolomite, and alumina, where all raw materials undergo high-temperature melting (1300–1600 °C) followed by fiberization using special metal bushings. GF production bears several environmental impacts including the extraction of raw materials (sand and rocks) that leads to land degradation, high energy consumption owing to high temperature melting and fiberization, CO_2 emissions from limestone decomposition and fuel combustion, and water and air pollution from the processing emissions and particulate matter. A comparative LCA among the two extensively used fibers demonstrates that GF production has a lower environmental impact than CF production but still significantly influences energy use and emissions.

Carbon fiber (CF) manufacturing involves converting polyacrylonitrile (PAN) or pitch-based precursors through stabilization (oxidation), carbonization (above 1000 °C), and surface treatment. CF production involves several environmental impacts including extremely high energy demand because of the carbonization stage, high utilization of processing chemicals such as solvents, surface treatment chemicals, and sizing agents, high CO_2, NO_x, and toxic waste emissions from PAN and CF manufacturing, and energy-production stages. CF production's energy demand is five times higher than that of GF manufacturing, making it one of the most resource-intensive components of FRP composites. Natural Fibers (NF) alternatives such as flax, jute, hemp, bamboo, etc. are making their way into FRP composites as sustainable alternatives to synthetic fibers. This growing interest can be credited to their bio-based origin. NF offers numerous environmental benefits including lower energy consumption compared to GF and CF, reduction in net CO_2 emissions owing to carbon sequestration during plant growth, biodegradability that lowers EoL's environmental impact, and reduced reliance on non-renewable fossil fuels. However, agricultural land use, water consumption, fertilizers, and pesticide use must be accounted for when evaluating their overall sustainability.

3.1.2 Environmental Impact of Resin Production

Polymeric resins play the role of matrix in FRP composites, bind fibers together, and provide structural integrity. The most used resins include epoxy, unsaturated polyester, and vinyl ester, all derived from fossil fuels. Epoxy resins are produced from epichlorohydrin and bisphenol-A, both are petrochemical derivatives. Epoxy resin manufacturing poses some serious environmental impacts including the depletion of fossil fuels (source of epoxy monomers), high-energy demand in chemical synthesis and polymerization, emission of volatile organic compounds as hazardous air pollutants, and toxicity concerns linked to bisphenol-A that poses risks both for the environment and human health. Polyester and vinyl ester resins are manufactured from styrene-based monomers that also originated from petroleum. Environmental impacts linked to the production of these resins are high VOC releases, styrene emissions, and energy consumption during polymerization and curing processes. Bio-based resins are making their way as alternatives to synthetic fossil fuel-based resins. These alternatives are reducing the reliance on non-renewable feedstock. Such bio-based resins include lignin-derived resins, soy-based epoxy, and furan-based thermosets. Renewable feedstocks, low carbon footprint, reduced GHG emissions, less toxicity and biodegradability are the potential benefits of these resins. However, high production costs, scalability challenges, and limited mechanical performance are downsides of these bio-based resin systems (Table 1).

Table 1 Comparative environmental impact of FRP raw materials

Environmental impact category	Glass fiber	Carbon fiber	Natural fiber	Epoxy resin	Polyester resin	Bio-based resin
Energy consumption	High	Very High	Low	High	Moderate	Low
CO_2 emissions	Moderate	High	Low	High	Moderate	Low
Toxicity	Low	Moderate	Low	High	High	Low
Fossil fuel dependency	High	High	None	High	High	Low
Biodegradability	None	None	High	None	None	Moderate

3.1.3 Environmental Impact of Additives and Fillers

Additives and fillers are incorporated in FRP composites to enhance their properties such as improving fire resistance, UV stability, and mechanical strength. However, their creation and integration into composites also lead to environmental concerns. Plasticizers and toughening agents are petroleum-based, therefore enhancing dependability on fossil fuels and posing hazards to the environment and life. UV light stabilizers such as hindered amine light stabilizers, when chemically degraded into side products, contribute to environmental pollution. Flame retardant additives are either phosphorus or brominated persist in the environment and are potentially toxic.

Fillers and reinforcements also present their environmental impact. For example, extracting and processing calcium carbonate and silica fillers require mining activities that lead to habitat destruction and dust emissions. Similarly, nanomaterials such as graphene and carbon nanotube synthesis are quite energy-intensive and present toxicity risks. Efforts are underway to reduce their environmental impact by developing bio-based or recycled fillers such as rice husk ash, wood fibers, and recycled CF.

3.1.4 Sustainable Strategies for Reducing Raw Material Impact

The environmental impact of FRP raw materials production can be alleviated by implementing several sustainability strategies. Some of them are previously discussed, such as increasing the use of recycled fibers, and resins to cut the demand for virgin materials, and the development of bio-based alternatives for both fibers (e.g., flax, hemp) and resins (e.g., lignin, soy-based). Other possibilities are improving energy efficiency in fiber and resin manufacturing through renewable energy integration and reducing hazardous chemical use in resin formulations and additive processing. The ultimate strategy is to be the implementation of circular economy models to promote waste reduction and closed-loop recycling.

3.2 Manufacturing and Processing Impacts (Energy Use, Emissions, Waste Generation).

The manufacturing and processing of FRP composites include substantial environmental impacts, primarily linked to high energy consumption, GHG emissions, and waste generation. FRP composite manufacturing typically involves fiber reinforcement processing, resin formulation, composite manufacturing (lay-up, moulding, filament winding, or pultrusion), and finishing processes. Each stage of FRP production contributes to the depletion of resources, emissions, effluents, and material waste, raising concerns about sustainability. The LCA approach assists in quantifying these impacts, directing the efforts to improve energy efficiency, reduce emissions, and enhance waste management. The subsequent section outlines the environmental footprint of FRP manufacturing, focusing on energy use, emissions, and waste generation, and discusses potential strategies for reducing the impact [6].

3.2.1 Energy Use in FRP Manufacturing and Processing

FRP composite manufacturing is energy-intensive, primarily because of high-temperature processes necessitated in fiber production, resin curing, and composite strengthening. The total energy consumption in the whole process depends on fiber type, resin system, and manufacturing method [4].

The production of GF involves heating silica-based raw materials to 1300–1600 °C, leading to considerable thermal energy consumption. The energy requirement of the process is estimated at 10–20 MJ/kg of GF produced. CF production is sizeably more energy-demanding than that of GF as it involves stabilization (about 300 °C), carbonization (above 1000 °C), and surface treatment. The energy demand of CF production can exceed 200 MJ/kg, making CF one of the most energy-intensive materials in FRP production. NF (flax, hemp, jute) production entails significantly lower energy input, principally for harvesting, retting, and fiber extraction. Their energy demand is assessed at 2–5 MJ/kg, making them a more sustainable alternative.

Epoxy, unsaturated polyester, and vinyl ester synthetic resins are essentially derived from petrochemicals, involving energy-intensive chemical synthesis and polymerization. The estimated energy demand to produce epoxy resin is 80–100 MJ/kg, while that of polyester resins is 40–60 MJ/kg. Bio-based resins are undoubtedly renewable, but their chemical processing and purification still demand substantial energy. However, the overall footprint of bio-based resin is lower than that of fossil-based synthetic resins.

Composite manufacturing is carried out through diverse techniques, requiring varying energy demands. Hand lay-up and vacuum bagging are the traditional methods and have moderate energy requirements but may cause high material waste. Compression molding and resin transfer molding require pressurization, and controlled heating, raising the electricity and thermal energy use. Pultrusion is continuous processing, that demands constant heating and drawing forces, rendering it an energy-intensive technique (30–60 MJ/kg of composite). Autoclave curing used for aerospace applications is the most energy-intensive process, as it requires both high pressure (up to 10 bar) and elevated temperatures (up to 180 °C) for several hours, leading to 500–1000 MJ/m^2 of composite panel.

3.2.2 Emissions from FRP Manufacturing and Processing

The production of FRP composites causes direct and indirect emissions, including GHG emissions (CO_2 equivalents), volatile organic compounds (VOCs), particulate matter (PM), and hazardous air pollutants (HAPs). CF manufacturing emits almost 5–30 kg CO_2-eq per kg of fiber, predominantly due to the high energy consumption in carbonization and precursor synthesis. GF manufacturing releases 0.5–2 kg CO_2-eq per kg of fiber, rising from fuel combustion and raw material processing. Resin synthesis also releases GHG emissions in significant amounts, for epoxy resin the emission is 3–6 kg CO_2-eq per kg of resin, and for polyester resins, it is about 2–4 kg CO_2-eq per kg. Composite manufacturing and thermal curing in ovens and autoclaves also contribute to CO_2 emissions from electricity and fuel combustion.

VOCs and hazardous air emissions are environmental pollutants and are strictly regulated in terms of exposure limits owing to potential health risks such as carcinogenicity. For example, the emission of styrene from polyester and vinyl ester resins during the curing process. Epoxy resins relatively emit lower VOCs but their processing releases amine-based hardeners and other toxic by-products. Particulate matter arises from the process of FRP cutting and machining (e.g., trimming, sanding, drilling). These processes generate fine dust and airborne fibers, causing inhalation hazards and environmental concerns.

3.2.3 Waste Generation in FRP Manufacturing

The manufacturing and processing of FRP composites generate various categories of waste, including fiber offcuts, resin-rich scraps, curing byproducts, and wasted tooling materials. It is estimated that 15–30% of total material use is often sent to landfills in the form of GF and CF offcuts due to limited recyclability. While NF waste is biodegradable and can be repurposed, hence alleviating the disposal concerns. Resin also appears as waste in the form of overflows, scrap, and excess materials. Hand lay-up and resin transfer molding produce scrap at a higher rate. Solvent-based hazardous liquid waste is yielded from cleansing agents such as acetone that are used in mold preparation and equipment maintenance. EoL management of composite waste requires a great deal of attention. Over 90% of the FRP composite waste presently ends up in landfills due to limited recycling ventures. Mechanical recycling involves grinding and reusing waste as fillers that reduce mechanical performance, so limited market acceptability. Thermal and chemical recycling are carried out through pyrolysis and solvolysis but require high energy input and specialized infrastructure.

3.2.4 Strategies for Reducing Environmental Impact

Industries can opt for certain sustainable strategies to mitigate the energy use, emissions, and waste associated with FRP manufacturing and processing, including improvements in energy efficiency, measures for emission control, and incorporating waste reduction and recycling innovations. For energy efficiency, industries can use renewable energy sources (solar, wind) for manufacturing, implement low-energy processing techniques (low-temperature curing resins, out-of-autoclave composites), and recycle the heat from high-temperature processes (e.g., carbon fiber pyrolysis). For emission control, industries can shift to low-VOC or solvent-free resins, implement styrene capturing and recovery systems for unsaturated polyester and vinyl polyester resin processing, and use high-efficiency filtration systems to limit particulate emission. Similarly, material waste can be reduced by opting for technologies like automation, precision cutting, and optimized lay-up techniques. There is a dire need for the development of closed-loop recycling processes for FRP scrap and explore bio-based and recyclable resins for improved EoL sustainability.

3.3 In-Use Phase: Durability Versus Sustainability Trade-Offs

FRP composites are valued for exceptional durability, high strength-to-weight ratio, corrosion resistance, and low maintenance requirements during the in-use stage. On one end, these characteristics contribute to longer service life and improved structural performance, while on the other end, they also present sustainability challenges. The principal trade-off in the in-use stage of FRP components is between durability (long-term performance, reduced material repair and replacement, and energy efficiency) and sustainability (EoL recyclability, material circularity, and environmental impact over time). This section explores the environmental impact of FRP composites during their in-use phase, focusing on how durability and performance advantages add to sustainability while discussing the challenges of recyclability, resource depletion, and waste accumulation at EoL [7].

3.3.1 Durability Benefits of FRP and Environmental Advantages

Durability is one of the key market features of FRP composites that nurture energy savings, lower GHG emissions, and reduce resource consumption during the in-use phase. Other key advantages are corrosion resistance, extended service life, lightweight nature, energy proficiency, low maintenance, and reduced material use. FRP composites, unlike metals and wood that degrade over time, are resistant to moisture, chemicals, and UV degradation. This resistance translates into longer product life and reduces the frequency of component replacement and materials consumption. Even in harsh industrial environments such as marine, FRP components can last over 50 years, significantly surpassing conventional materials. In the automotive and aerospace sectors, FRP composites reduce vehicle weight, resulting in lower fuel consumption in internal combustion engine-based vehicles, while extending battery life and driving range in electric vehicles. This consequence reduced GHG emissions due to lower energy demands. In wind energy applications, lighter turbine blades permit larger, and more efficient wind turbines, improving renewable energy output. FRP-based structures require nominal maintenance due to their high wear resistance, thus reducing the need for chemical surface treatments such as paints, anti-corrosion coatings, regular maintenance and replacements (saving materials, maintenance related energy and labor). In civil and structural engineering applications, such as bridges and building reinforcements, FRP decreases repair frequency, extending infrastructure lifespans with minor environmental costs [7].

3.3.2 Sustainability Challenges in the In-Use Phase

Although FRP's durability fetches environmental benefits, several sustainability concerns arise during its in-use phase, including EoL waste accumulation and management issues, embedded carbon and resource extraction concerns, components repairability, and materials circularity issues. In the first place, FRP's durability makes recycling and disposal challenging as they do not degrade easily. Secondly, the lack of large-scale recycling infrastructure results in the dumping of waste in landfills, resulting in increased waste accumulation and landfill pressure or incineration that causes high carbon emissions. The in-use phase of FRP reduces emissions, but high embodied energy during manufacturing could offset the balance between in-use phase energy savings and manufacturing emissions depending on the product's life. If the product is abandoned too early, the net environmental benefits could go negative. However, an FRP product well used over multiple decades could outweigh the initial environmental costs. FRP structures are difficult to repair, often demanding complete part replacement rather than localized maintenance. FRP has limited circular economy pathways, unlike metals that can be remelted and reused, making the reuse and reprocessing of FRP complex. Table 2 summarizes the trade-offs between durability and sustainability from a balanced perspective in FRP applications.

3.3.3 Strategies for Enhancing Sustainability in FRP's In-Use Phase

While taking maximum advantage of in-use phase durability, researchers and industries must tackle sustainability challenges through the development of recyclable and bio-based FRP, the adoption of circular economic practices, improvement in recycling technologies, and optimization through LCA-driven approaches. Bio-based resins derived from lignin, and soybean oil can condense embodied carbon emissions

Table 2 Durability versus sustainability trade-offs: a balanced perspective

Factor	Durability benefits	Sustainability challenges
Service life	Long-lasting materials reduce resource use	End-of-life disposal remains a challenge
Lightweight	Lower weight reduces energy use in transport	Embedded carbon from manufacturing remains high
Maintenance	Low maintenance reduces chemical treatments	Difficult to repair and refurbish
Recyclability	High strength prevents early failure	Recycling technologies are still inefficient
Material circularity	Extended use improves the environmental footprint	The lack of reuse options limits circular economy potential

while maintaining durability. Similarly, recyclable thermoplastic composites can enable easier EoL reprocessing. Circular economic practices such as remanufacturing and refurbishing can be implemented to extend FRP lifespan. Eco-design approaches such as design for disassembly can be explored to improve reusability. Similarly, manufacturers could be encouraged to optimize fiber volume fraction to reach a balance between performance and sustainability. Mechanical recycling processes can be implemented at large scale to repurpose the FRP scrap into secondary applications. Chemical recycling techniques (solvolysis, pyrolysis) require an advance to recover fibers and resins. LCA tools should be exploited to evaluate trade-offs between energy efficiency, material durability, and recyclability.

3.4 End-of-Life (EoL) Challenges: Disposal, Incineration, and Environmental Burden

EoL management remains a predominant sustainability challenge. FRP composites, unlike metals or glass, are difficult to recycle owing to their thermoset resin matrix, heterogeneous composition, and high durability, which makes disposal of them a critical issue. At present, most EoL FRP waste is either landfilled or incinerated, both of which have considerable environmental consequences. As discussed, landfilling contributes to waste accumulation and long degradation times, while incineration emits harmful releases and spends valuable raw materials. The lack of an efficient circular economy approach for FRP results in significant environmental burdens, including carbon emissions, resource depletion, and pollution risks. This section explores disposal, incineration, and the overall environmental impact of FRP at EoL, highlighting both challenges and potential solutions [8].

3.4.1 Landfilling of FRP Waste: Challenges and Environmental Impact

Landfilling is presently the dominant dumping route for FRP waste, with over 90% of composite waste ending up in landfills due to the lack of efficient recycling infrastructure. However, this method of disposing of the waste raises several concerns including space occupation, waste accumulation, long degradation time, leaching of hazardous additives, and microplastic creation. FRP composites are non-biodegradable and their degradation in nature spans centuries. This major setback leads to long-term waste accumulation in landfills. Landfilling spaces are becoming scarce as more and more waste is coming from the aerospace, automotive, construction, and wind energy industries. FRP contains toxic additives such as fire retardants, heavy metals, and styrene (a hazardous compound and constituent of ester resins) that would leach into soil and groundwater over time and cause long-term contamination problems. Over time, FRP will degrade into microplastic particles that could be transported through wind and water causing damage to both land and marine environments and ultimately contributing to ecosystem disruptions. Owing to these issues, there is an urgent need to cut landfill dependency by progressing alternative waste management strategies such as recycling, repurposing, or incineration with energy recovery.

3.4.2 Incineration of FRP Waste: Energy Recovery Versus Environmental Concerns

Incineration is the second most widely used methodology for managing FRP waste, particularly in regions with inadequate landfill space. It involves burning waste at high temperatures to generate energy, but it also comes with significant environmental setbacks including carbon emission, air pollution, and permanent loss of raw materials. However, the energy recovery potential of this method could be used to generate electricity. Combustion of GFRP and CFRP releases high levels of GHG including CO_2, CO, NO_x, and SO_x that deteriorate the air quality and contribute to climate change. Similarly, the burning of thermoset resins emits furans, dioxins, styrene, and other hazardous pollutants that are harmful to life on Earth. Incineration annihilates high-value fibers and resins into inert ash, inhibiting material recovery, and leading to resource depletion. The ashes are still in need to be landfilled. Energy recovery from waste requires high-efficiency combustion to generate electricity on technologically advanced incinerators. Although energy recovery partially offsets emissions, it does not offset at all the loss of primary materials, and the high carbon footprint of incineration. Therefore, Incineration should be considered a last resort after exploring repurposing and recycling options.

3.4.3 Overall Environmental Burden of EoL FRP Waste

The environmental burden of EoL FRP waste disposal surges from material loss, pollution risks, and emissions, making it one of the most challenging concerns of composite sustainability. As observed FRP composite production is an extremely energy-intensive activity, with poor EoL management further raising FRP lifecycle emissions rendering it a high carbon footprint product. The GHG emissions coming from landfilling and incineration condense the net environmental advantages gained from FRP's durability and lightweight properties during use. Inefficient EoL management of composite waste brings a lack of circular economic practices and resource inefficiency. Metals can be melted and reused indefinitely, but FRP materials used as alternatives to metal structures lost their structural integrity on discarding. Currently, less than 10% of FRP materials are recycled leading to massive raw materials loss and increased extraction of resources for both virgin fibers and resins. As highlighted earlier, both commonly used disposal strategies result in long-term environmental contamination risks such as microplastic formation, toxic leaching, and hazardous emissions for soil, water, and air. In marine applications (boats, offshore platforms, wind turbine blades), neglected FRP components add to ocean plastic pollution. The combination of high emissions, continual waste, and poor recyclability makes FRP waste a major sustainability challenge, requiring urgent technological and regulatory interventions.

3.4.4 Potential Solutions and Future Strategies for Sustainable EoL Management

Industries and policymakers must opt for innovative solutions, strict regulatory measures, and circular economy approaches to minimize the environmental burden of FRP disposal. Researchers and industries are required to innovate and adopt advanced recycling technologies and sustainable FRP alternatives. Mechanical recycling involves the shredding of the FRP wastes into small pieces for use as fillers in concrete, asphalt, and other composite applications. The major challenges linked to this recycling are the reduced mechanical properties of recovered materials and limited application for high-value products. Thermal recycling such as high-temperature pyrolysis recovers GF and CF while decomposing the resin matrix. The challenges linked to this technique are high energy demand and stringent requirements for emission control. Chemical recycling such as solvolysis involves the depolymerization under a solvent that allows dissolution of the resin matrix to separate the fibers and depolymerized products. The challenges associated with this approach are high costs, solvent recovery issues, and process scalability. On the other side, the development of bio-based resins such as ligning and vegetable oil-based polymers offer potential for biodegradability and improved recyclability. Thermoplastic resin-based composites can be melted and reformed, unlike thermosets, which are challenging to repurpose. Governments should implement strict regulatory measures such as enforcing landfill bans for FRP waste and encouraging recycling and reuse through incentives. EPR programs should be implemented, and industries should be willing to adopt them, where manufacturers are responsible for collecting and processing their waste at EoL. Circular economic models should be promoted where remanufacturing and repurposing should be encouraged rather than disposal and FRP waste should be reused in secondary application to extend the product life.

4 LCA of Different Recycling Methods for FRP Composites

FRP composites are extensively used across various industries due to their high strength-to-weight ratio, durability, corrosion resistance, and resistance to environmental degradation. However, their disposal at EoL offers significant environmental challenges owing to their non-biodegradability and hybrid and complex structures. Traditional waste management options, such as landfilling and incineration, are unsustainable due to the non-biodegradable nature of FRP and the release of harmful emissions during incineration as discussed in the preceding sections. As a result, various recycling techniques including mechanical, thermal, and chemical have been explored by researchers and industries as sustainable alternatives to traditional disposal methods. Mechanical recycling has emerged as a promising alternative for FRP waste management, offering a lower-energy recycling pathway compared to chemical and thermal methods. Thermal recycling processes, including pyrolysis, fluidized bed combustion, and gasification, offer the potential to recover high-value fibers and generate energy from FRP waste. Among these, chemical recycling has come out as a promising solution, offering the potential to reclaim high-quality fibers, and polymeric components for reuse. An LCA approach to any recycling approach provides a comprehensive evaluation of its environmental impact, energy efficiency, emissions profile, and sustainability compared to other recycling and disposal methods. This section explores the life cycle assessment (LCA) of mechanical, thermal, and chemical recycling of FRP composites, evaluating their environmental impact, benefits, challenges, and potential improvements to optimize their sustainability.

4.1 Mechanical Recycling: Grinding, Fiber Recovery, and Downcycling Impact

Mechanical recycling of FRP involves the fragmentation of composite materials into smaller particles through grinding, shredding, or milling, followed by their reuse as fillers or reinforcements in new composite or non-composite products. Mechanical recycling, unlike chemical and thermal recycling which allows the recovery of high-quality fiber, conserves the fiber-matrix bond but decreases fibers' length and mechanical characteristics. The retrieved material is mostly used in non-structural applications such as concrete reinforcement, asphalt additives, and low-grade composite manufacturing. The advantages of mechanical recycling include low energy consumption, minimal processing costs, and reduced GHG emissions in comparison to other recycling routes. However, issues such as fiber degradation, materials contamination, and limited high-value applications hamper its extensive acceptance. LCA is used to assess mechanical recycling's sustainability across its life cycle stages to understand its overall environmental footprint [9, 10].

4.1.1 Life Cycle Assessment Framework for Mechanical Recycling of FRP

LCA is a standardized scientific methodology used to estimate the environmental impacts of a process by analysing inputs (e.g., energy, raw materials, natural resources) and outputs (e.g., gaseous emissions, liquid effluents, solid residual waste) across its complete life cycle. The LCA of mechanical recycling of FRP usually follows four key stages in the way described as follows:

(a) *Goal and Scope Definition*

The objective in this stage is to evaluate the environmental benefits and setbacks of mechanical recycling in comparison to landfilling and incineration. Here, the system boundary starts with collection to end-use application, so the processes involved in LCA are collection, transportation, mechanical processing (grinding/milling), secondary material utilization, to end-use applications. Functional units (e.g., 1 ton of recycled FRP) are defined to standardize the impact assessment.

(b) *Life Cycle Inventory Analysis*

This stage involves data collection regarding energy consumption, machinery use, emissions from mechanical processing, and material outputs. Transportation emissions are also considered, particularly for centralized recycling facilities dealing in large FRP components such as wind turbine blades. The primary outputs with the quantification include recycled fiber-filled polymer blends, FRP powders, and potential secondary applications in the construction and manufacturing industry.

(c) *Life Cycle Impact Assessment*

Carbon footprint analysis estimates CO_2 emissions from mechanical processing compared to alternative disposal routes. Energy used and environmental impact are quantified, compared to chemical/thermal recycling, and virgin material production. Waste reduction potential assesses the percentage of FRP deterred from landfills and incinerators.

(d) *Interpretation and Improvement Strategies*

The outcomes help to recognize process inefficiencies, underline the areas for improvement, and indicate sustainable alternatives to improve mechanical recycling's viability.

4.1.2 Environmental Benefits of Mechanical Recycling

LCA studies on FRP mechanical recycling dependably exhibit substantial environmental advantages over landfilling and incineration processes. A comparative LCA of aerospace CFRP waste management in Europe found that mechanical recycling cut overall carbon emissions by 45% in comparison with landfill disposal [11]. Correspondingly, an LCA of wind turbine blade recycling in Germany established that using mechanically recycled FRP as secondary fillers in construction materials dropped resource depletion impacts by 30% [11]. Key environmental benefits of mechanical recycling include low energy consumption, reduced GHG emissions, and waste diversion. Mechanical recycling demands 50–70% less energy than that of chemical recycling, rendering it a more energy-efficient process. The release of CO_2, NO_x, and other harmful gases is avoided by preventing incineration and reducing the carbon footprint. Mechanical recycling mitigates landfill overflow and extends the material's life cycle by reinstating FRP waste into secondary applications.

4.1.3 Challenges and Limitations of Mechanical Recycling

Although mechanical recycling offers many advantages, yet it faces several limitations that restrict its large-scale adoption. The major challenges and limitations can be identified as fiber degradation, mechanical properties loss, limited high-value applications, processing and sorting complexities, and transportation and logistics issues. Grinding FRP composites into small particles shortens fiber length, abating its reinforcing capability in new applications. Mechanically recycled FRP is frequently used in low-grade, and non-structural applications rather than in high-performance composites, reducing its economic value. Variations in fiber type (GF, CF, NF), resin constitution, and contamination complicate the mechanical recycling process. The size of wind turbine blades, aircraft parts, and automotive panels makes the transportation of waste components to recycling centers and further adds to emissions and costs.

4.1.4 Strategies for Improving Mechanical Recycling Sustainability

Several strategies can be implemented such as decentralized recycling facilities, hybrid cycling approaches, eco-design for recyclability, and policy and economic incentives to enhance the environmental and economic feasibility of mechanical recycling. Local processing centers can be established to minimize transportation emissions and costs, making recycling more approachable. A hybrid approach such as mechanical in combination with chemical recycling (e.g. solvolysis assisted grinding) could aid in reducing the fibers' degradation and energy usage. The use of thermoplastic resins instead of thermosets could improve mechanical recycling and easy recovery of the materials. Governments and industries can support mechanical recycling through subsidies, landfill bans, and EPR program implementation [12].

In wrap-up, LCA studies of mechanical recycling of FRP composites highlight its environmental benefits, lower energy footprint, and potential for circular economy integration. However, issues such as fiber degradation, low-value applications, and transportation logistics must be tackled to improve its overall sustainability. The strategies highlighted previously will play a decisive role in making mechanical recycling a more feasible solution for composite waste management. Owing to rising environmental consciousness, industries are moving towards greener manufacturing and recycling practices. Mechanical recycling will remain a key player in reducing FRP waste, conserving resources, and minimalizing environmental impacts. Continued research, technological innovations, and supportive policies will be vital in ensuring that mechanical recycling becomes a scalable and efficient solution for FRP composite sustainability.

4.2 Thermal Recycling: Pyrolysis, Fluidized Bed Process, Energy Balance, and Emissions

Thermal recycling implies decomposing polymer matrices using heat to reclaim valuable fibers (e.g., CF or GF) while transferring the polymeric resin into useful byproducts such as syngas, oils, and char. The three main thermal recycling methods include pyrolysis, fluidized bed combustion, and gasification. In pyrolysis, FRP waste is heated in an oxygen-free environment to crack down the polymeric resin and reclaim fibers and gaseous byproducts. Gasification is the complete oxidation of FRP waste to produce synthetic gas (syngas) that is used as an energy source. Fluidized bed combustion involves the use of high-temperature air or sand beds to combust the resin, and leave behind clean, and reusable fibers. Thermal recycling provides high-quality fibers compared to mechanical recycling and is of suitable quality for reuse in structural applications. The shortcomings linked to thermal recycling are high energy-demanding and high gaseous emissions make it essential to conduct LCA studies to gauge the overall stability [9].

4.2.1 Life Cycle Assessment Framework for Thermal Recycling

LCA is a standardized methodology used to assess the environmental impact of a process from raw material input to EoL disposal. The LCA of thermal recycling of FRP follows similar four key stages in a way as depicted below:

(a) *Goal and Scope Definition*

Here, the objective is to measure the environmental performance of thermal recycling in comparison to mechanical recycling, landfilling, and incineration. The system boundary incorporates waste collection, transportation, thermal processing, energy recovery, fiber reuse, and final emissions management. The functional unit is typically chosen as 1 ton of recycled FRP waste, permitting comparison among different recycling methods.

(b) *Life Cycle Inventory Analysis*

This phase includes the collection of data regarding energy consumption, emissions from thermal decomposition, fiber recovery rates, and secondary product outputs. Transportation impacts are included, particularly for large-scale FRP waste such as discarded wind turbine blades, and aircraft parts. Key outputs that are estimated include recycled fibers, gaseous emissions (CO_2, NO_x, VOCs), and energy recovered either in the form of heat or syngas.

(c) *Life Cycle Impact Assessment*

Global warming potential is gauged through carbon footprint assessment by comparing CO_2 emissions from thermal recycling against other disposal approaches. The energy efficiency of the process is evaluated through the calculation of the net energy balance, where energy input for thermal processing and energy recovered as syngas are included. The air pollution index is determined by quantifying emissions of NO_x, CO, particulate matter, and VOCs. Waste reduction potential is assessed by estimating the percentage of FRP waste diverted from landfills through fiber recovery.

(d) *Interpretation and Recommendations*

The findings underscore the areas for improving thermal efficiency, emissions control, and secondary material applications. Complementary strategies such as renewable energy integration, and hybrid recycling approaches could be studied for better performance in various impact indicator categories.

4.2.2 Environmental Benefits of Thermal Recycling

LCA studies reveal that thermal recycling performs better than landfilling and incineration in terms of resource recovery and emissions reduction. Several case studies demonstrate its environmental benefits including fiber recovery for high-end applications, energy generation, and lower carbon footprint. An LCA analysis on aerospace CFRP waste in France established that pyrolysis recovered up to 80% of the original carbon fiber strength, making it appropriate for automotive, and industrial reuse [13]. An LCA study on wind turbine blade recycling in Denmark showed that gasification of FRP waste generated 15% more energy per ton compared to direct incineration, presenting it as a possible alternative to fossil fuels [14]. Interestingly, CF from pyrolysis-based recycling of CFPR composites causes 50–60% less CO_2 emissions than that of virgin fibers [15].

4.2.3 Challenges and Limitations of Thermal Recycling

In addition to its advantages, thermal recycling also presents several environmental and economic challenges including high energy demand, toxic emissions, air pollution, fiber recovery issues, and economic feasibility concerns. Pyrolysis and gasification demand significant energy input, reducing the overall sustainability of the processes unless driven by renewable energy sources. Nearly all combustion-based recycling methods release toxic gases such as VOCs, dioxins, and particulate matter that necessitate strict measures to control emissions. As a result of thermal recycling, CF retains good mechanical characteristics, but GF degrades at high temperatures and limits its reuse potential. The capital and operational costs of pyrolysis and fluidized bed reactors are very high which limits large-scale adoption without policy and economic incentives.

4.2.4 Strategies for Enhancing Sustainability of Thermal Recycling

The sustainability of thermal recycling can be improved by opting for several strategies including the integration with renewable energy sources, hybrid recycling approaches, advanced emission control technologies, and circular economy integration. The use of solar and wind-powered electricity can significantly offset the emissions from thermal recycling processes. Mechanical pre-processing could be integrated with thermal recycling which would optimize the fiber quality at one end and enhance the energy efficiency of the process on the other end. Carbon-capturing systems and gas-scrubbing technologies could be implemented to minimize the release of harmful pollutants. There is a need to develop a closed-loop supply chain in which reclaimed fibers would be reinstated in high-value applications to boost its economic viability.

In a long story short, the LCA study of thermal recycling of FRP composites underscores its potential as a sustainable waste management strategy, particularly for applications requiring high-quality fiber recovery and energy generation. Thermal recycling methods yield better properties of fibers compared to that of mechanical methods, making them suitable for reuse in high-end applications such as the automotive and aerospace industries. However, challenges regarding energy consumption, emissions, and cost-effectiveness must be tackled to warrant widespread adoption. Potential advancements in renewable-powered thermal processing, emission control technologies, and hybrid recycling systems will be crucial in making thermal recycling a scalable, efficient, and environment-friendly solution. As industries and policymakers are ambitious for zero-waste and circular economy models, thermal recycling will play a decisive role in handling FRP waste that lowers the reliance on virgin fiber production and minimizes the environmental impact of composites.

4.3 *Chemical Recycling: Solvolysis, Depolymerization, and Material Recovery Efficiency*

In chemical recycling, the polymer matrix is broken down using solvents, and catalysts. The fibers and monomers are separated and recovered from the mixture on the completion of chemical reactions. This method produces high-quality fibers while retaining their mechanical properties, making this approach superior to mechanical and even thermal recycling approaches. The three main chemical recycling processes for FRP composites are solvolysis, hydrolysis, and depolymerization. In solvolysis, the solvents such as alcohols, and acids are used to dissolve the resin matrix and reclaim the fibers with minimal damage. In hydrolysis, the water is used under acidic and alkaline conditions at high temperatures and pressures to break the polymeric resin and recover the fibers and other valuable compounds. Depolymerization is the process that converts the polymeric resin back into its original monomers and acts as a feedstock for new composite materials. Chemical recycling yields fibers with intact characteristics compared to mechanical and thermal recycling approaches. These fibers are more suitable for structural and high-value applications. However, chemical methods often require considerable energy and the use of hazardous chemicals, making it essential to perform the LCA study to assess the overall sustainability of the process [9].

4.3.1 Life Cycle Assessment Framework for Chemical Recycling

LCA is a systematic scientific method to estimate the environmental impacts of a process from raw material input to final product disposal. The LCA of chemical recycling of FRP follows four key stages in the manner as discussed here:

(a) *Goal and Scope Definition*

The objective of the LCA study is to evaluate the environmental performance of chemical recycling compared to mechanical recycling, thermal recycling, landfilling, and incineration. The system boundary for chemical recycling includes waste collection, transportation, chemical treatment, fiber recovery, secondary product utilization, emissions control, and final disposal. The functional unit is again typically 1 ton of chemically recycled FRP that allows for direct comparison across different recycling and disposal approaches.

(b) Life Cycle Inventory Analysis

In this phase, data is collected regarding chemical consumption, energy use, emissions, and fiber recovery efficiency. Transportation emissions are considered in the case of large FRP components. Outputs calculated include recovered fibers, chemical residues, emissions (e.g., GHG, VOCs), and secondary raw materials such as resins and monomers.

(c) *Life Cycle Impact Assessment*

The impact categories that need to be assessed here are carbon footprint, energy consumption, chemical waste, waste toxicity, and weight reduction potential. In carbon footprint analysis, CO_2 emissions from chemical recycling are to be compared with other EoL approaches. Energy intensity should be evaluated for solvolysis, hydrolysis, and depolymerization processes. There is a need to assess the environmental impact of solvent used and possible hazardous byproducts. In last, the quantification of the FRP materials is required that are successfully recovered in the form of fibers.

(d) *Interpretation and Optimization Strategies*

The conclusions emphasize areas for improvement such as the reduction in chemical use, optimized process efficiency, and integrated circular economy approaches.

4.3.2 Environmental Benefits of Chemical Recycling

LCA studies signify that chemical recycling offers substantial environmental benefits compared to landfilling and incineration. Several case studies highlight its advantages including high-quality fibers' recovery, reduction in the demand for virgin materials, and lower emissions of GHG. An LCA study on aircraft CFRP waste in Japan concluded that solvolysis recovered 95% of carbon fibers with minimal degradation, and made them suitable for reuse in new aerospace components [16]. A study on wind turbine blade recycling in Germany presented that chemical recycling decreased the need for virgin CF by 60%, lowering overall resource consumption [17]. Chemically recycled CFRP resulted in a 40–50% reduction in CO_2 emissions compared to virgin fiber production that has a high environmental footprint [18].

4.3.3 Challenges and Limitations of Chemical Recycling

Despite its environmental benefits, chemical recycling also encounters several limitations including high energy demand, high chemical demand, involvement of hazardous chemicals, cost, complexity, and large-scale implementation of the process. The process demands considerable heat and pressure that raises its energy footprint. Solvents and catalysts used in chemical recycling may be toxic, and volatile, and need appropriate waste management. Chemical recycling is more costly than mechanical recycling which limits its commercial adoption without encouraging financial incentives. Current chemical recycling facilities operate at a small or pilot-scale level that restricts their widespread use.

4.3.4 Strategies for Improving the Sustainability of Chemical Recycling

The sustainability of chemical recycling can be improved by opting for several strategies including utilization of green chemistry, integration of renewable energy, process optimization, and circular economic approaches. The use of biodegradable and non-toxic solvents can minimize environmental impact. Chemical recycling plants could be run at renewable energy resources to reduce emissions. With advancements in low-energy solvolysis and catalytical depolymerization, chemical process efficiency could be improved. By opting for the closed-loop supply chains where recovered materials including fibers and monomers can be reintegrated into new FRP products.

In short, the LCA study of chemical recycling of FRP composites emphasizes its potential as a sustainable and efficient waste management strategy compared to mechanical and thermal recycling. Chemical methods present top-quality fibers and facilitate polymer recovery making them suitable for high-end applications. There are several environmental advantages, adoption challenges, and strategies for improving the sustainability of the process as discussed. Chemical recycling will play a critical role in lowering FRP waste, minimizing environmental impacts, and supporting the transition to a sustainable composite industry.

4.4 Comparative LCA of These Methods to Assess Environmental Trade-Offs

A comparative LCA of these three recycling approaches delivers insights into their environmental trade-offs, energy efficiencies, emissions profiles, and resource recovery potentials. This section describes a systematic LCA comparison of mechanical, thermal, and chemical recycling, emphasizing their relative advantages, limitations, and optimization strategies to enhance sustainability. The three primary recycling approaches differ in terms of process efficiency, material recovery, energy demand, and environmental impact which have already been discussed in the preceding sections [19].

4.4.1 Life Cycle Assessment Framework

LCA is a standardized process used to estimate the environmental impacts of different recycling methods throughout their lifecycle. The comparative LCA of mechanical, thermal, and chemical recycling follows these key stages:

(a) *Goal and Scope Definition*

The objective is to compare the environmental performances of mechanical, thermal, and chemical recycling of FRP composites. The system boundary starts with waste collection and ends at secondary product utilization through intermediate process including transportation, recycling processes, energy consumption, emissions, and fiber recovery. The functional unit is as usual 1 ton of recycled FRP, permitting direct comparison across different methodologies.

(b) *Life Cycle Inventory Analysis*

Table 3 A comparison of impact categories of three different recycling processes

Impact category	Mechanical recycling	Thermal recycling	Chemical recycling
GWP	Lowest GWP due to minimal energy requirements	Moderate GWP due to emissions from polymer decomposition	Highest GWP due to energy-intensive solvent processing
Energy consumption	Low energy demand (grinding and processing)	High energy input for pyrolysis and combustion	Very high energy demand for solvent-based depolymerization
Air pollution and emissions	Generates minimal airborne emissions but produces fine dust	Produces CO_2, NO_x, and VOCs, requiring emission controls	Produces solvent vapors and wastewater requiring treatment
Waste reduction and quality of recovered materials	Yields lower-quality fibers with limited reuse options	Produces energy and high-purity fibers but involves polymer loss	Best fiber quality and potential for polymer recovery

Energy input, emissions, fiber recovery rates, waste residues, chemical use, and secondary material production are the key inputs whose quantification is required. Data regarding the outputs linked to the recycling technique under study. For mechanical recycling are recycled fiber-reinforced fillers, energy consumption for grinding, dust, and particle emissions. For thermal recycling are recovered fibers, gaseous emissions (CO_2, NO_x, VOCs), and energy produced from polymer decomposition. For chemical recycling are high-purity fibers, chemical byproducts, solvent residues, and wastewater treatment requirements.

(d) *Life Cycle Impact Assessment*

Environmental impact categories to be analyzed and compared in this LCA study are global warming potential (CO_2 Emissions), energy consumption, air pollution and emissions, and waste reduction and quality of recovered materials. The impacts are presented in Table 3.

(d) *Interpretation (Trade-Offs)*

The outcomes need to be interpreted in terms of comparative environmental trade-offs.

4.4.2 Comparative Environmental Trade-Offs

The LCA results underscore meaningful trade-offs between the three recycling methods as illustrated in Table 4. Mechanical recycling is the most energy-efficient and environmentally friendly method but limits fiber reuse to low-value applications. Thermal recycling offers better fiber quality and energy recovery but has a higher carbon footprint due to emissions. Chemical recycling generates high-purity fibers

Table 4 Environmental trade-offs of recycling methods

Impact category	Mechanical recycling	Thermal recycling	Chemical recycling
Energy consumption	Low	High	Very high
Carbon footprint	Lowest	Moderate	Highest
Fiber quality	Poor	Moderate	High
Polymer recovery	None	Limited	High
Emissions (CO_2, NO_x, VOCs)	Low	High	Moderate
Economic feasibility	High	Moderate	Low

and polymer recovery, which makes it the best option for circular economy strategies, but it comes with high energy and chemical usage.

In short, the comparative LCA of mechanical, thermal, and chemical recycling reveals that each method has distinct benefits and trade-offs. While mechanical recycling is energy-efficient and low-impact, it results in fibers' quality loss. Thermal recycling recovers high-quality fibers and energy but emits pollutants, while chemical recycling achieves the highest fiber and polymer recovery but at significant energy and chemical costs. A multi-method, circular economy-driven approach will be needed for sustainable FRP recycling, lowering environmental impact, and fostering resource efficiency.

5 Case Studies: LCA of Recycled FRP Composites

5.1 Automotive Sector: Recycled CFRP/GFRP in Lightweight Structures.

The automotive industry is increasingly using carbon fiber-reinforced polymers (CFRP) and glass fiber-reinforced polymers (GFRP) to manufacture weight-reduced vehicles that boost fuel efficiency and reduce emissions. The recyclability of these materials is crucial to their sustainability [20]. Recycling CFRP and GFRP is challenging due to the intractable separation of fibers from resin, loss of material properties, and high recycling expenses. Some of these challenges in recycling CFRP and GFRP are being addressed using various processes such as mechanical recycling (grinding to smaller particle sizes to reuse), thermal recycling (pyrolysis to retrieve fibers at the cost of degrading the resin), and chemical recycling (solvolysis to yield high-purity fibers) [21].

One of the most highly praised examples of CFRP recycling in automotive is that of BMW i3, one of the first commercial electric cars to extensively use CFRP in its body structure [22]. BMW developed a closed-loop recycling process to retrieve fibers from waste during manufacturing and end-of-life components. By using pyrolysis, fibers were separated and used in load-bearing components such as interior trims

Table 5 A comparison of CFPR and GFRP recycling case studies at two different automotive manufacturers

Parameter	Recycled CFRP (BMW i3)	Recycled GFRP (Ford)
Recycling method	Pyrolysis	Mechanical recycling
Main application	Body panels, interiors	Underbody shields, brackets
CO_2 reduction	40% versus virgin CFRP	20% versus virgin GFRP
Cost savings	Moderate	High
Structural use	Limited	Moderate

and roofs. Life cycle analysis (LCA) studies found that CFRP used in BMW i3 led to a 40% decrease in CO_2 emission compared to virgin CFRP, minimized landfill waste, and conserved resources. Although CFRP recycling is energy-intensive, its long-term benefits such as decreased vehicle emissions during use justify its initial burden on the environment [11].

Similarly, Ford was also in the position to effectively apply recycled GFRP in vehicles such as Ford Escape and Fusion. Ford used mechanical recycling instead of thermal or chemical recycling, grinding GFRP waste to a powdery texture and mixing it with virgin resin. The process was used to produce underbody shields, brackets, and non-structural reinforcements that maintained their mechanical strength without causing harm to the environment. Compared to virgin GFRP, the application of recycled GFRP resulted in a 20% reduction in manufacturing energy consumption and a 30% reduction in material costs, making it a low-cost, environmentally friendly option. In a comparison of the two approaches, the BMW i3's application of recycled CFRP was focused on high-performance applications but was expensive and limited to non-structural parts. In Ford's vehicles, on the other hand, recycled GFRP was cheaper, applicable in a wide range of applications, and easy to process, though lower in strength compared to CFRP. Despite such progress, there are still challenges such as high recycling costs, inconsistency in material properties, and regulations that limit their application. However, innovations in hybrid recycling technologies, AI optimization, and more cross-industry cooperation are making a more circular and sustainable approach to composite material application in cars a reality [21] (Table 5).

5.2 *Wind Energy: Repurposing Turbine Blades Versus Landfill Disposal.*

Wind energy is growing rapidly as a renewable power source, but the disposal of fiber-reinforced polymer (FRP) composite turbine blades remains a major environmental challenge. Most modern wind turbine blades are made from glass fiber-reinforced polymers (GFRP) or carbon fiber-reinforced polymers (CFRP), designed to be durable and lightweight. However, when these blades reach the end of their 20–25-year lifespan, their disposal becomes problematic [23]. Traditional recycling methods struggle to separate fibers from polymer matrices, making landfill disposal the most common solution. Unfortunately, this approach is unsustainable because blades take up significant landfill space, do not biodegrade, and contribute to long-term environmental damage. Regulations in the European Union and parts of the U.S. are now restricting landfill use for FRP materials, pushing the industry to explore alternative disposal methods. One promising solution is repurposing decommissioned turbine blades for secondary applications, extending their useful life and reducing environmental impact [24]. A key initiative in this area is the Re-Wind Network, which explores innovative ways to reuse turbine blades instead of discarding them. In Ireland and the U.S., researchers have successfully repurposed old blades into pedestrian bridges, bike shelters, and noise barriers for highways [25]. A bridge built in Cork, Ireland, incorporated turbine blades as structural components, demonstrating that these composite materials can be effectively reused for durable infrastructure. Similarly, in Denmark, decommissioned blades have been transformed into playground structures, utilizing their strength and resistance to harsh weather conditions.

Life cycle assessment (LCA) studies comparing repurposing with landfill disposal reveal that repurposing reduces CO_2 emissions by up to 30–50% while also having a significantly lower energy footprint compared to recycling through thermal or chemical processes. Since repurposing requires minimal processing, it is a cost-effective alternative that keeps materials in use for longer. However, it is not a complete solution—logistics, transportation costs, and adaptation to new structural applications remain challenges, and repurposing alone cannot handle the increasing volume of decommissioned blades. A combination of chemical recycling, circular economy strategies, and policy incentives will be needed to manage turbine blade waste effectively [26]. Innovations in chemical recycling could allow for high-quality fiber recovery, while designing new turbine blades for easier recyclability will ensure a more sustainable future. Government policies and industry collaboration will also play a crucial role in promoting the development of environmentally friendly materials and closed-loop recycling systems [27]. Repurposing wind turbine blades is a practical and immediate alternative to landfill disposal, offering environmental benefits while supporting sustainable infrastructure projects. However, as wind farms continue to expand, the industry must invest in scalable recycling technologies and circular design strategies to ensure a truly sustainable wind energy future.

5.3 Construction Industry: Reuse and Downcycling Applications.

The construction industry is one of the largest consumers of materials and produces significant amounts of waste, making sustainability a serious concern. FRP composites, frequently used in bridges, building facades, and reinforcing structures, offer high strength, corrosion resistance, and durability. However, their EoL disposal presents challenges, as conventional recycling methods struggle to separate fibers from polymer matrices. As a result, the industry is exploring two key approaches to managing FRP waste sustainably: reuse and downcycling. Reuse extends the lifespan of composite materials in similar or new applications. Downcycling repurposes them into lower-value products, averting them from becoming landfill waste. Several case studies demonstrate how these strategies are being implemented successfully in construction. One of the most notable examples of reuse in construction comes from the bridge infrastructure sector. In the Netherlands, a project repurposed decommissioned FRP bridge panels into modular pedestrian bridges and temporary walkways [28]. The panels, originally designed for high loads and environmental resistance, retained sufficient mechanical properties even after their initial service life. Instead of dumping them, engineers cut and reshaped the panels to fit new structural requirements, demonstrating that FRP components can be directly reused in non-critical load applications with minimal processing. The reuse strategy not only extended the life of the material but also reduced the carbon footprint associated with manufacturing new composite panels.

Another successful example of reuse is observed in the adaptive reuse of FRP facade panels from decommissioned buildings. In the United Kingdom, an office building undergoing renovation recovered its FRP curtain wall panels and repurposed them for façade cladding in the new construction. The panels were cleaned, modified, and integrated into the design of a modern energy-efficient structure, sustaining their aesthetic and functional properties. This approach minimized destruction waste, reduced demand for virgin materials, and aligned with circular economy principles in architecture. While direct reuse is ideal, not all FRP waste can be used in its original form due to wear and structural limitations. This is where downcycling becomes a possible alternative [29]. One of the most impactful downcycling applications has been seen in road and pavement construction. In Germany, shredded FRP waste from decommissioned wind turbine blades and demolition projects has been successfully incorporated into asphalt mixtures for road surfaces [30]. The composite fibers improve crack resistance and durability, enhancing road performance while reducing dependence on virgin aggregates. LCA studies on this project indicated that using FRP-reinforced asphalt lowered overall emissions by 25% compared to conventional materials, proving that downcycling can offer both environmental and economic benefits.

Another noteworthy downcycling initiative comes from concrete reinforcement applications. Traditionally, steel reinforcement is used in concrete structures, but researchers in Japan have investigated using recycled FRP fibers from old boats and automotive parts to strengthen precast concrete blocks. The mechanically processed FRP fibers when mixed with concrete mixtures, improved tensile strength and impact resistance, making them appropriate for non-load-bearing walls, pavements, and noise barriers. This approach not only switches composite waste from landfills but also reduces the corrosion issues associated with traditional steel reinforcements. Despite the success of reuse and downcycling strategies, challenges persist. The heterogeneous nature of FRP composites, including differences in resin types and fiber content, makes it difficult to establish standardized processing methods. Additionally, logistics and transportation costs for repurposing large FRP components can be appreciable. However, with the increasing regulatory burden to limit landfill disposal and growing advancements in recycling technologies, the construction industry is probably to adopt more scalable and efficient FRP reuse and downcycling solutions [30].

In conclusion, the reuse and downcycling of FRP composites in construction present practical solutions to the growing issue of composite waste management. From repurposing bridge panels and facade elements to incorporating recycled FRP fibers in roads and concrete, these strategies demonstrate how the industry can move forward to a more sustainable and circular economy. While challenges remain, ongoing research and industry collaborations will play a crucial role in scaling up these efforts and making FRP recycling an integral part of sustainable construction practices.

5.4 Lessons Learned from LCA Studies on Real-World Recycling Projects

LCA is a robust tool used to evaluate the environmental impacts of recycling processes that help industries make learned decisions about sustainability. Over the period, LCA studies on real-world recycling projects have provided valuable insights into material recovery, energy consumption, emissions reduction, and overall environmental trade-offs. These lessons are transforming industries' approach to recycling strategies, circular economy principles, and policy development. One of the key lessons learned from LCA studies is that the choice of recycling method significantly impacts the overall environmental impact. Mechanical recycling, which involves grinding and reusing fibers in new applications, often requires low energy input but results in materials with reduced mechanical properties. In contrast, thermal and chemical recycling methods, such as pyrolysis and solvolysis, can recover higher-quality fibers but demand substantial energy, leading to increased GHG emissions unless powered by renewable sources. For instance, an LCA study on wind turbine blade recycling in Germany revealed that mechanical recycling had a 30% lower carbon footprint than pyrolysis, but the recovered fibers were only suitable for low-load applications like insulation panels rather than structural components [31]. This finding highlights the trade-offs between energy use and material quality in recycling choices.

Another important lesson from LCA studies is that closed-loop recycling systems offer greater environmental benefits than open-loop systems. Closed-loop recycling refers to recovering materials and reintegrating them into the same or similar high-performance applications, reducing the demand for virgin resources. A case study on CFRP recycling in the aerospace sector demonstrated that closed-loop recycling of carbon fibers resulted in a 50% reduction in raw material extraction impacts, compared to open-loop recycling where fibers were used in lower-grade applications like automotive interiors. However, the challenge with closed-loop recycling lies in preserving fiber quality and achieving economic feasibility, which requires continued innovation in fiber extraction technologies.

Furthermore, LCA studies emphasize the importance of integrating renewable energy sources into recycling processes to mitigate emissions. Many chemical and thermal recycling methods consume high amounts of electricity and heat, which, if sourced from fossil fuels, can negate the environmental benefits of recycling. An LCA comparison of FRP waste processing in the European Union found that using solar or wind-powered energy in pyrolysis reduced overall emissions by 40%, making the process far more sustainable. This underscores the need for industries to align recycling efforts with renewable energy adoption for maximum environmental gains. Another significant insight from LCA studies is that transportation logistics play a crucial role in the overall sustainability of recycling. Large FRP components, such as wind turbine blades and aircraft fuselages, are difficult to transport due to their size and weight. An LCA study on blade recycling in the U.S. showed that transporting decommissioned blades to centralized recycling facilities contributed up to 25% of the total carbon footprint of the process. In contrast, on-site or decentralized recycling approaches, where materials are processed closer to their disposal sites, proved to be more sustainable and cost-effective. This lesson highlights the importance of regional recycling infrastructure to minimize emissions from material transportation. LCA studies also reveal that designing products with recyclability in mind from the start (eco-design) is crucial for long-term sustainability. At present, many FRP composites are not designed for continent recycling that leads to high energy and resource requirements for material recovery. Researchers in Japan found that CFRP recycling in automotive becomes far more efficient, reducing energy use by 35% and increasing material recovery rates using thermoplastic-based composites instead of thermoset resin matrices. This emphasizes the need for industries to adopt circular economy principles in material selection and product design to accelerate future recycling.

LCA findings warrant that policy, and economic incentives are necessary to drive wider adoption of recycling practices. While recycling FRP composites is environmentally beneficial, it is often more expensive than landfill disposal or incineration EoL approaches, leading many industries to vote for the least sustainable option. However, LCA-driven policy directives, such as landfill bans in the EU and EPR programs, have pushed manufacturers to invest in sustainable recycling technologies. The case of Denmark's wind turbine blade recycling program demonstrates how government policies, coupled with economic incentives, led to a 60% surge in recycled blade materials within five years. This demonstrates that sustainability-driven policies can substantially accelerate the transition toward circular material use.

In a wrap-up, LCA studies on real-world recycling projects present essential insights into the environmental trade-offs, energy requirements, and economic challenges linked to different recycling approaches. They emphasize the significance of selecting the right recycling method, prioritizing closed-loop systems, integrating renewable energy, optimizing logistics, and designing materials for recyclability. Additionally, they emphasize the role of policy support and economic incentives in making recycling both practical and scalable. To leap forward, continuous advancements in recycling technologies, supply chain optimization, and regulatory frameworks will be decisive in transforming FRP recycling into a truly sustainable and cost-effective solution for industries globally.

6 Advancements in Sustainable FRP Composites

In realizing sustainable FRP composites, advancements are rapidly morphing the materials landscape by combining high-performance attributes with environmental responsibility. A most emerging development is the use of thermoplastic-based FRP composites. Thermoplastic matrices can be melted and reshaped, unlike traditional thermoset matrices that are irreversibly cured, which not only simplifies processing and shortens cycle times but also enables recycling and remanufacturing. This feature significantly improves the composite's sustainability profile by permitting closed-loop recycling and reducing waste at EoL. Recent research also has focused on developing bio-based resins that would substitute traditional petrochemical matrices with renewable choices derived from lignin, soy, and other natural sources. These bio-based resins not only cut the carbon footprint related to composite production but also boost biodegradability, paving the way for eco-friendly EoL disposal options. Beside resin innovations, there has been a significant trend to integrating NF—such as hemp, flax, jute, and basalt—in place of conventional synthetic reinforcers. NF offers comparable mechanical properties while staying renewable, cost-effective, and possessing lower embodied energy, making them ideal for sustainable composite applications. In addition to material replacement, surface functionalization, and modifications have appeared as key advancements to enhance fiber-matrix adhesion. Techniques including alkaline treatment, silane coupling, and graft copolymerization improve the interfacial bonding between NF and polymer matrices, thereby minimizing moisture uptake and increasing durability. These improvements are crucial for extending the service life of FRP composites and ensuring their performance in challenging applications [5].

Equivalently, research into smart and self-healing composites is bringing promising results. By introducing microcapsules or vascular networks encompassing healing agents into the composite, these materials can autonomously restore the damages such as microcracks. Self-healing FRPs not only extend the service life but also moderate the need for recurrent repairs and replacements, leading to lower overall material consumption and maintenance costs. This is a significant benefit for sustainable infrastructure and high-performance applications. The modular composite design signifies another transformative advancement. By developing standardized, exchangeable composite modules, manufacturers can assemble larger structures more efficiently while ensuring that damaged modules can be easily replaced or upgraded without discarding the entire component. This design approach promotes efficient resource use and supports the circular economy by enabling the reuse and repurposing of FRP components, thus minimalizing waste and environmental burden linked to EoL disposal. Advanced manufacturing techniques are further improving the sustainability of FRP composites. Methods such as tailored fiber placement and additive manufacturing sanction for precise control over fiber orientation and placement, minimizing material waste and enabling the production of components that are optimized for load bearing and efficiency. These technologies facilitate the creation of complicated geometries with minimal excess material, reducing energy consumption during production and enabling rapid prototyping as well as scalable production runs. The integration of nanotechnology, incorporating nanoparticles such as graphene or carbon nanotubes, has led to significant improvements in mechanical strength, thermal stability, and barrier attributes. Nanofillers not only reinforce the composite structure but also add to enhanced recyclability by improving the compatibility between recycled and virgin materials [32].

Recycling and circular economy initiatives have also gained a thrust. Conventional recycling methods including mechanical grinding, thermal pyrolysis, and chemical recycling are being developed to optimize material recovery while minimalizing energy consumption and emissions. Closed-loop recycling systems are emerging as a feasible solution to reduce waste and dependence on virgin materials. This approach not only preserves resources but also reduces the environmental burden of composite discard. Digital tools and machine learning algorithms have started playing a transformative role in composite design and production. These technologies promote a more comprehensive LCA and allow informed decision-making by optimizing process parameters and forecasting performance outcomes. Enhanced data availability and standardized LCA methodologies are vital for pinpointing environmental hotspots and directing sustainable design choices. All these advancements signal a promising future for sustainable FRP composites, where high-performance materials are obtained to reduce environmental impact throughout their life cycle. As research keeps pushing the boundaries of material science, collaborations among academia, industry, and policymakers will be indispensable in opting for these innovations and warranting that FRP composites contribute effectively to a circular and sustainable materials economy.

7 Challenges, Recommendations, and Future Directions

The sustainability of FRP composites depends on effective recycling strategies supported by the LCA studies. Various recycling techniques, environmental impacts, and challenges related to composites' EoL management have been surveyed in detail. Significant measures are required across the research, industry, and policymaking to improve the sustainability of FRP recycling. LCA should be adopted as a decision-making tool for assessing the environmental trade-offs and sustainability in FRP recycling. However, the data gaps, methodological irregularities, and lack of standardized assessment criteria as examined previously limit the widespread implementation of LCA and are regarded as primary barriers. LCA-driven decision-making can be more impactful by tackling these deficiencies in data availability and standardization issues. In the first place, it is necessary to develop comprehensive and standardized databases for LCA of FRP materials by designated various fiber-resin combinations. The databases should also be featured with the use of diverse manufacturing and fabrication processes and regional variations. A close collaboration among academia, industry, and governmental bodies will warrant transparent data-sharing to make LCA studies more dependable. The system's boundaries also need to be carefully defined to confirm consistent comparison among the different recycling technologies. Both global and regional databases should be expanded to be incorporated with thorough datasets on materials life cycles, FRP production, environmental discharges, service life, EoL scenarios, and recycling efficiencies. Standardization of LCA methodologies for composites would allow reliable cross-study comparison. Techno-economic analysis (TEA) is integrated with LCA to depict the holistic view of both environmental and financial sustainability. Material flow models should be designed to measure the long-term stabilities of diverse recycling approaches. Despite the rising environmental consciousness regarding the importance of FRP composite recycling, several economic and policy-related barriers hinder large-scale adoption. At present, recycling approaches especially chemical and thermal bear high costs that restrict commercial viability. On the other side, the market acceptability of recycled FRP materials is limited due to poor mechanical characteristics and comparatively high processing costs. Weak regulatory directives are unable to oblige composite recycling or boost sustainable alternatives. The lack of financial incentives and subsidies discourages the industry from investing in recycling projects.

All the stakeholders including researchers, industrialists, and policymakers are required to put their inputs to converge their efforts in realizing the sustainable future of composites recycling. Researchers and scientists should divert their attention to advancing chemical recycling techniques to reclaim the fibers and resin using minimum energy and chemicals. A long-term solution lies in materials innovations and eco-friendly design. They should strive to foster alternative sustainable fibers, and bio-based and recyclable resins. Bio-based sustainable materials including natural fibers and bio-based resins lower the dependability on non-renewable petroleum resources and decompose back to nature [5]. Recyclable resin allows the efficient and easy recovery of fibers from composites. Thermoplastic-based resins can remelted multiple times and allow many time reuses in composites in contrast to traditional thermoset resins. Hybrid recycling approaches using combinations of different technologies must be researched to realize higher efficiency. LCA methodologies should be substantiated with dynamic models to forecast the long-term environmental impact of FRP composite recycling. Researchers should keep their emphasis on the recyclability of the FRP composite parallel to advancement in composites. The composite industries must shift their efforts to realize circular economic practices that are vital for the survival of mankind on the planet. The industry should close its loop by opting for the best practices that one way or another support recycling. Integration of recovered fibers and resins into new composites will lessen the burden on the exploitation of natural resources that are not renewable. Industries should design their FRP products by following recyclability principles so that they can be efficiently disassembled and processed at EoL. Adoption of automated separating technologies will allow the industry the proficient recovery of materials and minimize the labor-intensive recycling steps. The industry should work in symbiosis that would develop a mechanism for the use of recycled materials in secondary applications such as transportation, construction, and energy segments. Strong policy invention is the need of the hour to address all the challenges linked to the composite recycling industry. Government institutions and policymakers must implement ERP directives to incentivize manufacturers to take responsibility for their composite wastes. There is a need to provide financial incentives and subsidies for the industries investing in recycling technologies and set regulatory targets for the composite industry like those established for the metal and plastic industries. FRP should be standardized as the waste class for proper collection and sorting to facilitate recycling. The governments should offer funds for research projects on circularity and sustainability in composite materials.

In a nutshell, LCA plays a pivotal role in channeling sustainable decision-making for FRP recycling by assessing environmental impacts and identifying the most sustainable EoL pathways. However, its effectiveness depends on improving data availability, standardizing methodologies, and integrating economic feasibility assessments. Future research should focus on improving the accuracy, accessibility, and applicability of LCA studies in composite materials. Sustainable FRP recycling can evolve from a niche practice to a mainstream circular economy solution by implementing technological innovations, regulatory support, and industry-wide collaboration. With consistent efforts in eco-friendly and sustainable material development, optimized recycling procedures, and comprehensive sustainability assessments, FRP composites can contribute to a greener and more sustainable future. [33]

8 Conclusion

The sustainability of FRP composites is a rising concern owing to their augmenting use in the automotive, aerospace, wind energy, and construction industries. While these materials offer extraordinary durability, lightweight properties, and outstanding mechanical performances, but their complicated composition and the lack of natural degradation pose major environmental setbacks—especially at the EoL stage. This chapter offers an in-depth evaluation of sustainable recycling approaches using LCA as a systematic tool, investigating the challenges, opportunities, and trade-offs linked to various recycling methods. One of the key takeaways is that FRP composites demand a well-versed EoL strategy to minimize landfill waste and maximize material recovery. Mechanical recycling is a simple and cost-effective approach. However, fiber degradation and downcycling outcomes limit its long-term efficacy. Thermal recycling methods including pyrolysis and combustion, can retrieve both energy and fibers, but these processes are energy-demanding and can release higher emissions. Chemical recycling appears to be a promising strategy to recover fibers and resins in almost in their original quality, but its high processing costs and dependence on chemicals for processing deter its widespread adoption. The comparative LCA of these approaches features the need to balance environmental, economic, and technological factors while deciding the most sustainable solution for different applications.

The principal challenge in this field is the non-existence of standardized LCA methodologies for FRP composites. Variations in fiber types, resin compositions, and manufacturing processes make it challenging to generalize environmental impact assessments. Furthermore, the long service life of these composite materials and indecisive disposal pathways further complicate sustainability modelling. Enriched LCA databases, standardized system boundaries, and improved impact assessment tools are crucial for more precise evaluations. Real-time case studies from the automotive, wind energy, and construction industries demonstrated both the potential and troubles of FRP recycling. In the automotive sector, recycled CF and GF can substantially reduce the weight of vehicles and carbon footprint, enhancing fuel efficiency, and lowering emissions. The wind energy industry is facing the challenge of turbine blade disposal and exploring methods to repurpose and advance recycling practices to reduce landfill waste. In the construction industry, recycled FRP composites can be repurposed for non-structural applications and secondary utilization. These case studies point out the importance of advanced recycling strategies, material repurposing, and policy-driven sustainability efforts. On the other side, innovations in recyclable resin systems, bio-based fibers, and smart material designs will play a decisive role in improving FRP sustainability. Governments and industries must develop a collaboration to conclude stronger regulations, ERP programs, and financial incentives that boost responsible FRP waste management. Circular economy principles such as aiming on reuse, remanufacturing, and high-value recycling, should be embedded throughout the complete FRP life cycle, from the design phase to the disposal stage.

In short, while there are momentous challenges to overcome, a sustainable future for FRP recycling is possible through technological innovations, regulatory support, and industry-wide commitment. By considering a holistic approach that balances environmental impacts with economic feasibility, FRP composites can evolve from being a waste-intensive industry to a key player in a circular and sustainable economy.

References

1. Memon H, Wei Y, Zhu C (2022) Recyclable and reformable epoxy resins based on dynamic covalent bonds—present, past, and future. Polym Test 105:107420
2. Sbahieh S, Mckay G, Al-Ghamdi SG (2023) A comparative life cycle assessment of fiber-reinforced polymers as a sustainable reinforcement option in concrete beams. Original Res 9. (in English)

3. Ziemińska-Stolarska A, Sobulska M, Pietrzak M, Zbiciński I (2024) Application of life cycle assessment to analysis of fibre composite manufacturing technologies in shipyards industry. 12(3):461
4. Song YS, Youn JR, Gutowski TG (2009) Life cycle energy analysis of fiber-reinforced composites. Compos Part A: Appl Sci Manuf 40(8):1257–1265
5. Sahari J, Sapuan SM, El-Shekeil YA, Ishak MR, Akhtar R (2015) Natural fibre-reinforced thermoplastic starch composites. In: V PM, Yu L (eds) Starch-based blends, composites and nanocomposites. The Royal Society of Chemistry, p 0
6. McIlhagger A, McIlhagger R (2020) Manufacturing processes for composite materials and components for aerospace applications, pp 59–81
7. Lunetto V, Galati M, Settineri L, Iuliano L (2023) Sustainability in the manufacturing of composite materials: a literature review and directions for future research. J Manuf Process 85:858–874
8. Ziemińska-Stolarska A, Sobulska M, Pietrzak M, Zbiciński I (2024) A review of end-of-life scenarios for fibre-reinforced polymer materials. 17(15):3713
9. Veres C, Tănase M (2025) A bibliometric and systematic review of the use of recycled composite materials with an emphasis on the mechanical performance of structures. 18(3):607
10. Makenji K (2010) 9—mechanical methods for recycling waste composites. In: Goodship V (ed) Management, recycling and reuse of waste composites. Woodhead Publishing, pp 217–252
11. Witik RA, Teuscher R, Michaud V, Ludwig C, Månson J-AE (2013) Carbon fibre reinforced composite waste: An environmental assessment of recycling, energy recovery and landfilling. Compos Part A: Appl Sci Manuf 49:89–99
12. Gupt Y, Sahay S (2015) Review of extended producer responsibility: a case study approach. 33(7):595–611
13. Petrakli F et al (2020) End-of-life recycling options of (nano)enhanced CFRP composite prototypes waste-a life cycle perspective. Polymers (Basel) 12(9). (in eng)
14. Anu et al (2020) Multifarious pretreatment strategies for the lignocellulosic substrates for the generation of renewable and sustainable biofuels: a review. Renew Energy 160:1228–1252
15. Ciacci L, Zattini G, Tosi C, Berti B, Passarini F, Giorgini L (2022) Carbon fibers waste recovery via pyro-gasification: semi-industrial pilot plant testing and LCA. 14(7):3744
16. Okajima I, Sako T (2017) Recycling of carbon fiber-reinforced plastic using supercritical and subcritical fluids. J Mater Cycles Waste Manage 19(1):15–20
17. Cong N, Song Y, Zhang M, Wu W (2023) Life cycle assessment of carbon reduction potential of EoL wind turbine blades disposal scenarios in China. Environ Impact Assess Rev 100:107072
18. Diez-Cañamero B, Mendoza JMF (2023) Circular economy performance and carbon footprint of wind turbine blade waste management alternatives. Waste Manage 164:94–105
19. Meng F, McKechnie J, Turner T, Wong KH, Pickering SJ (2017) Environmental aspects of use of recycled carbon fiber composites in automotive applications. Environ Sci Technol 51(21):12727–12736
20. Kretschmer J (1988) Composites in automotive applications–state of the art and prospects. Mater Sci Technol 4(9):757–767
21. La Rosa AD, Banatao DR, Pastine SJ, Latteri A, Cicala G (2016) Recycling treatment of carbon fibre/epoxy composites: materials recovery and characterization and environmental impacts through life cycle assessment. Compos Part B: Eng 104:17–25
22. Kretschmer J (2013) Composites in automotive applications–state of the art and prospects. Mater Sci Technol 4(9):757–767
23. Liu P, Barlow CY (2017) Wind turbine blade waste in 2050. Waste Manage 62:229–240
24. Karavida S, Nõmmik RJAUA (2015) Denmark. Waste management of end-of-service wind turbines
25. Ruane K et al (2023) Construction and cost analysis of bladebridges made from decommissioned FRP wind turbine blades. 15(4):3366
26. Jani HK, Singh Kachhwaha S, Nagababu G, Das A (2022) A brief review on recycling and reuse of wind turbine blade materials. Mater Today: Proc 62:7124–7130

27. Heng H, Meng F, McKechnie J (2021) Wind turbine blade wastes and the environmental impacts in Canada. Waste Manage 133:59–70
28. Speksnijder S (2018) Reuse of wind turbine blades in a slow traffic bridge
29. Ziegler L, Gonzalez E, Rubert T, Smolka U, Melero JJJR, Reviews SE (2018) Lifetime extension of onshore wind turbines: a review covering Germany, Spain, Denmark, and the UK. 82:1261–1271
30. Mishnaevsky L (2021) Sustainable end-of-life management of wind turbine blades: overview of current and coming solutions. 14(5):1124
31. Nagle AJ, Mullally G, Leahy PG, Dunphy NP (2022) Life cycle assessment of the use of decommissioned wind blades in second life applications. J Environ Manage 302:113994
32. Parveez B, Kittur MI, Badruddin IA, Kamangar S, Hussien M, Umarfarooq MA (2022) Scientific advancements in composite materials for aircraft applications: a review. 14(22):5007
33. Deeney P et al (2021) End-of-life alternatives for wind turbine blades: Sustainability Indices based on the UN sustainable development goals. Resourc Conserv Recycl 171:105642

MIX
Papier aus verantwortungsvollen Quellen
Paper from responsible sources
FSC® C105338

If you have any concerns about our products,
you can contact us on
ProductSafety@springernature.com

In case Publisher is established outside the EU,
the EU authorized representative is:
Springer Nature Customer Service Center GmbH
Europaplatz 3, 69115 Heidelberg, Germany

Printed by Libri Plureos GmbH
in Hamburg, Germany